CARTESIAN VECTORS AND TENSORS

Ram Bilas Misra

CWP

Central West Publishing

Related Titles

Glossary of Mathematical Terms and Concepts (Part III)
ISBN (print): 978-1-925823-73-8

Glossary of Mathematical Terms and Concepts (Part IV)
ISBN (print): 978-1-925823-74-5

CARTESIAN VECTORS AND TENSORS

by

Prof. Dr. Ram Bilas Misra

Ex Vice Chancellor, Avadh University, Faizabad, U.P. (India);
Professor of Mathematics, Research & Strategic Studies Centre,
Lebanese French University, Erbil, Kurdistan (Iraq).

Former: *Dean*, Faculty of Science, A.P. Singh University, Rewa, M.P. (**India**);
Prof., Dept. of Maths., Higher College of Edn., Aden Univ., Aden (**Yemen**);
Professor & Head, Dept. of Maths. & Stats., A.P.S. University, Rewa, M.P. (**India**);
Prof., Dept. of Maths., College of Science, Salahaddin University, Erbil (**Iraq**);
UGC Visiting Prof., Mahatma Gandhi Kashi *Vidyapith*, Varanasi, U.P. (**India**);
Professor, Dept. of Maths, Ahmadu Bello Univ., Zaria (**Nigeria**) – designate;
Prof. & Head, Dept. of Maths. & Comp. Sci., Univ. of Asmara, Asmara (**Eritrea**);
Director, Unique Inst. of Business & Technol., Modi Nagar, Ghaziabad, U.P. (**India**);
Prof. & Head, Dept. of Maths., Phys. & Stats., Univ. of Guyana, Georgetown (**Guyana**);
Prof. & Head, Dept. of Maths., Eritrea Inst. of Technology, Mai Nefhi (**Eritrea**);
Prof.& Head, Dept. of Maths., School of Engg., Amity Univ., Lucknow, U.P. (**India**);
Prof. & Head, Dept. of Maths. & Comp. Sci., PNG Univ. of Technology, Lae (**PNG**);
Prof. of Maths., Teerthankar Mahaveer University, Moradabad, U.P. (**India**);
Prof., Dept. of Maths, Oduduwa Univ., Ipetumodu, Osun State (**Nigeria**) – designate;
Prof., Dept. of Maths, Adama Science & Technology Univ., Adama (**Ethiopia**);
Prof. & Head, Dept. of Maths. & C.S., Bougainville Inst. of Bus. & Tech., Buka (**PNG**) – designate;
Prof. & Head, Dept. of Maths., J.J.T. University, Jhunjhunu, Rajasthan (**India**);
Dean, Faculty of Science, J.J.T. University, Jhunjhunu, Rajasthan (**India**);
Professor, Dept. of Maths, Wollo University, Dessie, Wollo (**Ethiopia**);
Professor, Dept. of Appld. Maths., State Univ. of New York, Incheon (**S. Korea**);
Prof., Dept. of Maths. & Computing Sci., Divine Word Univ., Madang (**PNG**);
Director, Maths., School of Sci. & Engg., Univ. of Kurdistan Hewler, Erbil (**Iraq**);
DAAD Fellow, University of Bonn, Bonn (**Germany**);
Visiting Professor, University of Turin, Turin (**Italy**);
Visiting Professor, University of Trieste, Trieste (**Italy**);
Visiting Professor, University of Padua, Padua (**Italy**);
Visiting Professor, International Centre for Theoretical Physics, Trieste (**Italy**);
Visiting Professor, University of Wroclaw, Wroclaw (**Poland**);
Visiting Professor, University of Sopron, Sopron (**Hungary**);
Reader, Dept. of Maths. & Stats., South Gujarat University, Surat, Gujarat (**India**);
Reader, Dept. of Maths. & Stats., University of Allahabad, Allahabad, U.P. (**India**);
Asst. Prof., Dept. of Maths., College of Sci., Mosul Univ., Mosul (**Iraq**) – designate;
Senior most *NCC Officer* (Naval Wing), Univ. of Allahabad, Allahabad, U.P. (**India**);
Lecturer, Dept. of Maths., KKV Degree College, Lucknow, U.P. (**India**).

2020

This edition has been published by Central West Publishing, Australia

For more information about the books published by Central West Publishing, please visit https://centralwestpublishing.com

Disclaimer
Every effort has been made by the publisher and author while preparing this book; however, no warranties are made regarding the accuracy and completeness of the content. The publisher and author disclaim without any limitation all warranties as well as any implied warranties about sales, along with fitness of the content for a particular purpose. Citation of any website and other information sources does not mean any endorsement from the publisher and author. For ascertaining the suitability of the contents contained herein for a particular lab or commercial use, consultation with the subject expert is needed. In addition, while using the information and methods contained herein, the practitioners and researchers need to be mindful for their own safety, along with the safety of others, including the professional parties and premises for which they have professional responsibility. To the fullest extent of law, the publisher and author are not liable in all circumstances (special, incidental, and consequential) for any injury and/or damage to persons and property, along with any potential loss of profit and other commercial damages due to the use of any methods, products, guidelines, procedures contained in the material herein.

NATIONAL LIBRARY OF AUSTRALIA

A catalogue record for this book is available from the National Library of Australia

ISBN (print): 978-1-925823-82-0

DEDICATED TO

MY ELDERS AND WELL-WISHERS

ESPECIALLY

Shri Mishree Gir, Semrai, Dist. Lakhimpur-Kheri
(for his monetary help in our hard days);

Shri Ramadhar Bajpeyi, Semrai
(for his monetary help in our hard days);

Shri Devi Dayal Pandey, Semrai
(for his unconditional paternal love and support to my family);

Shri Rameshwar Dayal Pandey, Semrai
(for his life-long support in my all endeavours);

Pt. Jagannth Prasad Pandey - a freedom fighter of Semrai
(for his sacrifices, modesty and simplicity);

Shri Chhail Bihari Awasthi - the first graduate of Semrai
(for his wisdom and multi-dimenisonal personality);

Shri Babu Ram Awasthi
(Teacher, Public Intermediate College, Gola Gokarannath);

Shri Gulab Gir, Gola Gokarannath
(for his generosity and affection);

Shri Ram Avtar Tiwari, Gola Gokarannath
(my eldest maternal uncle – for his love and appreciation);

Shri Sadho Ram Shukla, Bhikhampur
(a freedom fighter helping me to get admitted in B.Sc. class);

Shri Girija Dayal Trivedi, Lucknow
(for my admission in B.Sc. class & for his generosity);

Dr. Rishi Keshav Pandey, IAS, Lucknow
(for his unforgettable cooperation in hard times);

Shri Surendra Nath Shukla, Gola Gokarannath
(a trustworthy classmate from Class VI to XII).

CONTENTS

PREFACE

Idea of writing this book has roots in an invitation to the author extended by National Institute of Technology, Dimapur, Nagaland (India) to deliver lectures in a Workshop for the M.Sc. (Physics) students of the Institute during October 29 – November 6, 2019. The material in the book covers the topics needed for a course in Cartesian Vectors and Tensors with applications to Geometry and Theory of Relativity.

The subject matter is presented here in six chapters of which the first one deals with the vector algebra. Derivation of vector-valued functions is considered in Chapter 2. The Chapter 3 briefly mentions about the vectors in electric and magnetic fields. The next two chapters are devoted to the detailed discussion of Cartesian Tensors. Chapter 5 includes the Tensors in cylindrical and spherical coordinates. The last chapter presents a brief introduction of Theory of Relativity.

Chapters are divided into Sections and the discussion within the Sections is presented in the form of Definitions, Theorems, Corollaries, Notes and Examples. Most of the chapters end in a Problem Set containing unsolved exercises, but solution to challenging ones is provided with necessary hints. The sub-titles within the Sections are numbered in decimal pattern. For instance, the equation number (c.s.e) refers to the eth equation in the sth section of Chapter *c*. When *c* coincides with the chapter at hand, it is dropped. Adequate references to the results appeared earlier are made in the text avoiding their unnecessary repetition. Double slashes marked at the end of Theorems, Corollaries, Solutions of Exercises, etc., indicate their completion.

The author's long teaching career of more than ***five decades*** at various universities round the globe and research expertise in different fields helped him for lucid presentation of the subject. Perhaps it is the divine will that the organizers of the course provided this opportunity to the author to offer the homage to his Ph.D. supervisor (late) Professor Dr. R. S. Mishra at University of Allahabad, Prayagraj (India) in his centenary year. The book is dedicated to his teachers and other mentors. Thanks are also due to various Universities all over the world especially University of Allahabad (India); University of Guyana, Georgetown (Guyana); P.N.G. University of Technology, Lae (Papua New Guinea); Adama Science & Technology University, Adama (Ethiopia); State University of New York, Incheon (South Korea); Divine

Word University, Madang (P.N.G.), etc., where I gained a lot while exposing my expertise. Organizers of the Workshop also deserve my gratitude. It may go unfair on my part if I do not record the sincere co-operation of my family especially the better-half (Mrs. Rekha Misra), whom I often had to exhaust for my academic passion. During preparation of the book, I lost the last cousin of my father (Shri Parmeshwar Deen Misra), a retired teacher, who was younger to me for ten months. I pray for his eternal peace and enough strength to the bereaved family and friends to mourn his irreparable loss forever. Sincere thanks are also due to the publisher for their valuable cooperation and bringing the book into limelight in a limited time.

Although proofs are read with utmost care and solutions to problems are verified repeatedly, yet an oversight or any discrepancy brought to the notice of the author by the inquisitive readers(s) shall be thankfully acknowledged.

Lucknow (India): October 19, 2019 Ram Bilas Misra

CHAPTER 1

VECTOR ALGEBRA

§ 1. Physical quantities

Fig. 1.1

In our discussion, we shall mainly encounter with two types of quantities: scalars and vectors. Those having only magnitude are called *scalars*. For instance, area, density, length, mass, speed, volume, numbers, etc., are all scalars. On the other hand, the physical quantities equipped with both magnitude and direction are called *vectors*. Acceleration, displacement, electric field, force, magnetic field, velocity, weight, etc. are vectors.

Let us consider a line OP of length, say a. The directed line $\overline{\text{OP}}$ represents a vector (from O to P). It may be denoted by a bold face letter, say **a**, while a scalar will be written in *italic* letter. The length a of the line OP is called the *magnitude* of the vector **a**. In symbols, we denote it by $|\mathbf{a}| = a$.

Note 1.1. A vector with zero magnitude is called a *zero* or *null vector*. It is denoted by **0**. Its direction is *indeterminate*.

1.1. Addition of vectors

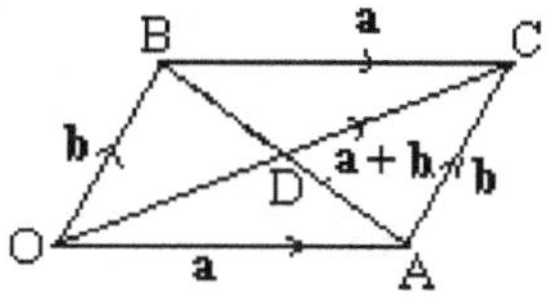

Fig. 1.2

Let $\overline{\text{OA}} = \mathbf{a}$ and $\overline{\text{OB}} = \mathbf{b}$ be two vectors acting at a point O. Complete the parallelogram OACB and draw its diagonals OC and BA. The vector along the diagonal $\overline{\text{OC}}$ is defined as the *vector sum* of the vectors $\overline{\text{OA}}$ and $\overline{\text{OB}}$. The opposite sides OB and AC are equal and parallel; and the vectors along them are in same sense of direction. So they are equal. We, therefore, write it as

$$\overline{\text{OC}} = \overline{\text{OA}} + \overline{\text{AC}} = \overline{\text{OA}} + \overline{\text{OB}} = \mathbf{a} + \mathbf{b}. \quad (1.1)$$

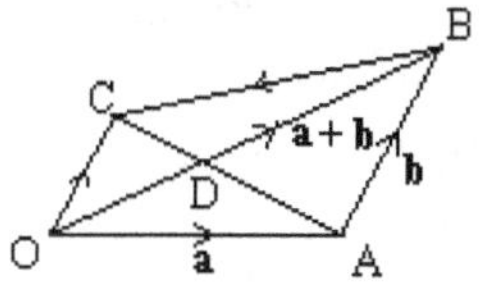

Fig. 1.3

Similarly, $\overline{\text{OA}} = \overline{\text{BC}} = \mathbf{a}$. The vector along the other diagonal BA is defined as the *difference* of two vectors $\overline{\text{OA}}$ and $\overline{\text{OB}}$:

$$\overline{\text{BA}} = \overline{\text{OA}} - \overline{\text{OB}} = \mathbf{a} - \mathbf{b}. \quad (1.2)$$

Theorem 1.1. The vector sum is commutative and associative:

$$\mathbf{a} + \mathbf{b} = \mathbf{b} + \mathbf{a}, \tag{1.3}$$

$$\mathbf{a} + (\mathbf{b} + \mathbf{c}) = (\mathbf{a} + \mathbf{b}) + \mathbf{c} \tag{1.4}$$

Proof. (i) Cf. Fig.1.2. We have

$$\overline{OC} = \overline{OB} + \overline{BC} = \overline{OB} + \overline{OA} = \mathbf{b} + \mathbf{a},$$

which, in association with Eq. (1.1), establishes Eq. (1.3).

(ii) In Fig.1.3, LHS of Eq. (1.4) = $\mathbf{a} + \overline{AC} = \overline{OC}$. Also, its

$$\text{RHS} = \overline{OB} + \mathbf{c} = \overline{OC}. \ //$$

Corollary 1.1. For every vector **a**, there hold $\mathbf{a} + \mathbf{0} = \mathbf{0} + \mathbf{a} = \mathbf{a}$.

Example 1.1. Three vectors **a, b, c** act along the consecutive sides of a triangle. Show that their vector sum vanishes. Generalize the result for any closed polygon.

Solution. Let the vectors form a triangle ABC where $\overline{AB} = \mathbf{a}$, $\overline{BC} = \mathbf{b}$ and $\overline{CA} = \mathbf{c}$. By definition of vector sum,

$$\overline{AB} + \overline{BC} = \overline{AC}, \qquad \text{i.e.} \qquad \mathbf{a} + \mathbf{b} = -\mathbf{c}$$

$$\Rightarrow \qquad \mathbf{a} + \mathbf{b} + \mathbf{c} = \mathbf{c} + (-\mathbf{c}) = \mathbf{0}.$$

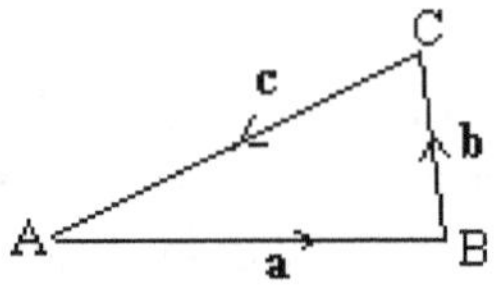

Fig. 1.4

Generalization can be similarly proved. //

Example 1.2. ABCD is a parallelogram with G as the point of intersection of its diagonals. For any point O prove that

$$\overline{OA} + \overline{OB} + \overline{OC} + \overline{OD} = 4\overline{OG}.$$

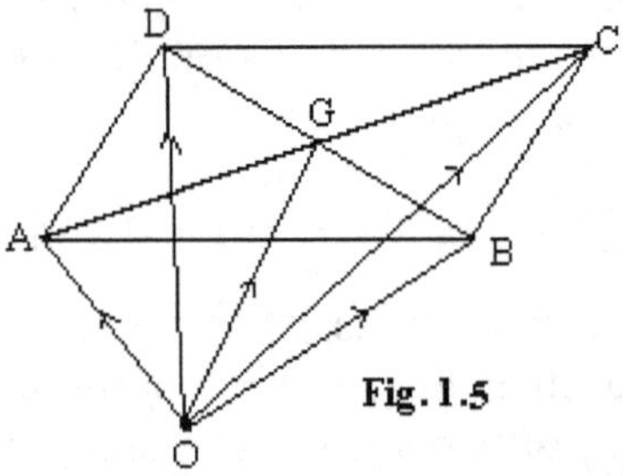

Fig. 1.5

Solution. The vector $\overline{OG}$ can be expressed as the sum of vectors:

$$\overline{OG} = \overline{OA} + \overline{AG}, \ \overline{OG} = \overline{OB} + \overline{BG}, \ \overline{OG} = \overline{OC} + \overline{CG} \text{ and } \overline{OG} = \overline{OD} + \overline{DG}.$$

Adding these vectors, we get

$$4\,\overline{OG} = (\overline{OA} + \overline{OB} + \overline{OC} + \overline{OD}) + (\overline{AG} + \overline{BG} + \overline{CG} + \overline{DG}).$$

The diagonals of a parallelogram bisect each other. Thus, G being the midpoint of both diagonals, we have

$$\overline{AG} + \overline{CG} = \overline{BG} + \overline{DG} = \mathbf{0}.$$

This reduces above result to the desired form.

§ 2. Scalar multiplication of a vector

Let x be a scalar and $\mathbf{a}$ some vector then $x\mathbf{a}$ is also a vector with magnitude $|\,x\,|$ times that of $\mathbf{a}$. The direction of $x\mathbf{a}$ will be along (or opposite to) $\mathbf{a}$ if x is positive (or negative).

Definition 2.1. Two vectors $\mathbf{a}$ and $\mathbf{b}$ are called *collinear* iff one of them is some scalar multiple of the other:

$$\mathbf{a} = t\,\mathbf{b}, \tag{2.1}$$

where $t \neq 0$ is some scalar. The magnitudes of collinear vectors are proportional. If, in addition, their magnitudes are same above vectors are called *equal* (or *opposite*) when $t = 1$ (or -1).

Theorem 2.1. Scalar multiplication of vector(s) satisfies the following laws:

(*Associative law*)
$$x\,(y\,\mathbf{a}) = (x\,y)\,\mathbf{a}, \tag{2.2}$$

(*Distributive laws*)

$$(x + y)\,\mathbf{a} = x\,\mathbf{a} + y\,\mathbf{a},\ \ x\,(\mathbf{a} + \mathbf{b}) = x\,\mathbf{a} + x\,\mathbf{b}. \tag{2.3}$$

Proof being simple is left for the reader. //

Definition 2.2. A vector with unit magnitude is called a *unit vector.*

The unit vector along any non-null vector $\mathbf{a}$ is obtained by multiplying the vector by the reciprocal of its magnitude:

$$\hat{\mathbf{a}} \equiv \mathbf{a}\,/\,|\,\mathbf{a}\,|\,. \tag{2.4}$$

Example 2.1. Prove the following results geometrically:

$$\mathbf{a} = (\mathbf{a} + \mathbf{b})/2 + (\mathbf{a} - \mathbf{b})/2, \tag{2.5}$$

$$\mathbf{b} = (\mathbf{a} + \mathbf{b})/2 - (\mathbf{a} - \mathbf{b})/2. \tag{2.6}$$

Solution. Cf. Fig. 1.2. The diagonals OC and BA of the parallelogram bisect each other in the point D. Therefore,

$$\overline{OD} = (1/2)\,\overline{OC} = (\mathbf{a} + \mathbf{b})/2 \qquad \text{and} \qquad \overline{DA} = (1/2)\,\overline{BA} = (\mathbf{a} - \mathbf{b})/2.$$

Their sum determines the vector $\overline{OA} = \mathbf{a}$. This establishes Eq. (2.5).

Next, in the Δ OBD: $\overline{OB} = \overline{OD} - \overline{BD} = (1/2)\,\overline{OC} - (1/2)\,\overline{BA}$ giving Eq. (2.6). //

Example 2.2. Show that the sum of vectors along the medians (directed from the vertices) of a triangle vanishes.

Solution. Let AD, BE and CF be the medians of the triangle ABC. The vectors $\overline{BD}$ and $\overline{BC}$ are in same sense of direction and the former is of half magnitude to that of the latter. Hence,

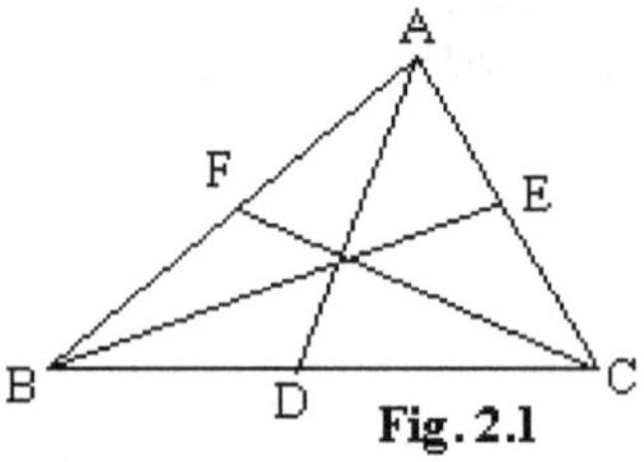

Fig. 2.1

$$\overline{BD} = (1/2)\,\overline{BC} \Rightarrow \overline{AD} = \overline{AB} + \overline{BD} = \overline{AB} + (1/2)\,\overline{BC}.$$

Similarly,

$$\overline{BE} = \overline{BC} + (1/2)\,\overline{CA} \qquad \text{and} \qquad \overline{CF} = \overline{CA} + (1/2)\,\overline{AB}.$$

Adding these vectors, by Example 1.1, we get

$$\overline{AD} + \overline{BE} + \overline{CF} = (3/2)\,(\overline{BC} + \overline{CA} + \overline{AB}) = \mathbf{0}. //$$

§ 3. Components of a vector along the coordinate axes

Let P (x, y, z) be any point in the space E_3 with position vector $\overline{OP} =$ **r** with respect to some origin O and three rectangular coordinate axes Ox, Oy, Oz through O. Let us draw the perpendicular PL from P to the plane xOy. Perpendiculars from L to the coordinate axes Ox and Oy are also drawn. If $\hat{\mathbf{i}}, \hat{\mathbf{j}}, \hat{\mathbf{k}}$ be

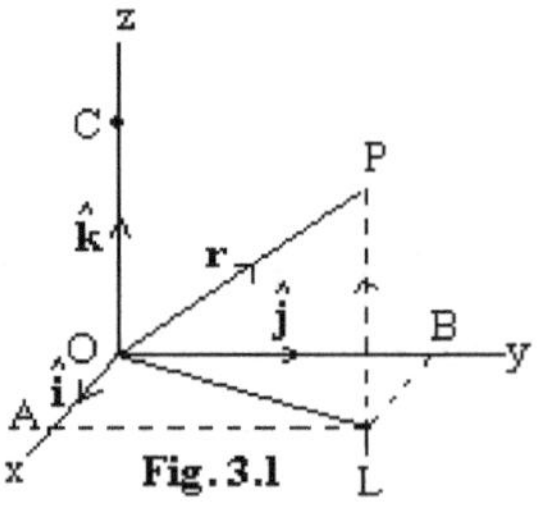

Fig. 3.1

the unit vectors along the respective coordinate axes then,

$$\overline{\text{OA}} = (\text{OA})\,\hat{\mathbf{i}} = x\,\hat{\mathbf{i}}, \quad \overline{\text{OB}} = (\text{OB})\,\hat{\mathbf{j}} = y\,\hat{\mathbf{j}}.$$

Similarly, $\overline{\text{OC}} = \overline{\text{LP}} = z\hat{\mathbf{k}}$, where C is taken on O$z$ line at a distance equal to z from O. By the definition of vector addition, we have

$$\overline{\text{OA}} + \overline{\text{AL}} = \overline{\text{OL}} \quad \text{and} \quad \overline{\text{OL}} + \overline{\text{LP}} = \overline{\text{OP}}.$$

Therefore,

$$\overline{\text{OP}} \equiv \mathbf{r} = \overline{\text{OA}} + \overline{\text{AL}} + \overline{\text{LP}} = \overline{\text{OA}} + \overline{\text{OB}} + \overline{\text{OC}}$$

$$= x\,\hat{\mathbf{i}} + y\,\hat{\mathbf{j}} + z\hat{\mathbf{k}} \equiv (x, y, z)\,. \tag{3.1}$$

Definition 3.1. The numbers x, y, z are called the *components* of the vector $\overline{\text{OP}}$ along the rectangular coordinate axes Ox, Oy, Oz respectively. They also determine the coordinates of the point P. In particular,

$$\left.\begin{array}{l} \mathbf{0} = (0, 0, 0) = 0\,\hat{\mathbf{i}} + 0\,\hat{\mathbf{j}} + 0\,\hat{\mathbf{k}}, \;\; \hat{\mathbf{i}} = (1, 0, 0) = 1\hat{\mathbf{i}} + 0\,\hat{\mathbf{j}} + 0\,\hat{\mathbf{k}}, \\ \hat{\mathbf{j}} = (0, 1, 0) = 0\,\hat{\mathbf{i}} + 1\hat{\mathbf{j}} + 0\,\hat{\mathbf{k}}, \;\; \hat{\mathbf{k}} = (0, 0, 1) = 0\,\hat{\mathbf{i}} + 0\,\hat{\mathbf{j}} + 1\,\hat{\mathbf{k}}. \end{array}\right\} \tag{3.2}$$

Theorem 3.1. The following vectors are equal if $a_i = b_i$, $i = 1, 2, 3$:

$$\mathbf{a} = (a_1, a_2, a_3) \quad \text{and} \quad \mathbf{b} = (b_1, b_2, b_3)\,. \tag{3.3}$$

Proof. The vectors

$$\mathbf{a} = a_1\,\hat{\mathbf{i}} + a_2\,\hat{\mathbf{j}} + a_3\,\hat{\mathbf{k}} \;\text{ and }\; \mathbf{b} = b_1\,\hat{\mathbf{i}} + b_2\,\hat{\mathbf{j}} + b_3\,\hat{\mathbf{k}} \tag{3.4}$$

are equal if

$$(a_1 - b_1)\,\hat{\mathbf{i}} + (a_2 - b_2)\,\hat{\mathbf{j}} + (a_3 - b_3)\,\hat{\mathbf{k}} = \mathbf{0} = (0, 0, 0)$$

$\Rightarrow$

$$a_1 - b_1 = a_2 - b_2 = a_3 - b_3 = 0 \Rightarrow a_1 = b_1,\; a_2 = b_2,\; a_3 = b_3.$$

Conversely,

$$a_1 = b_1,\; a_2 = b_2,\; a_3 = b_3 \Rightarrow a_1\,\hat{\mathbf{i}} = b_1\,\hat{\mathbf{i}},\; a_2\,\hat{\mathbf{j}} = b_2\,\hat{\mathbf{j}},\; a_3\,\hat{k} = b_3\,\hat{\mathbf{k}}.$$

Their addition determines

$$a_1\,\hat{\mathbf{i}} + a_2\,\hat{\mathbf{j}} + a_3\,\hat{\mathbf{k}} = b_1\,\hat{\mathbf{i}} + b_2\,\hat{\mathbf{j}} + b_3\,\hat{\mathbf{k}}, \quad \text{i.e.} \quad \mathbf{a} = \mathbf{b}.$$

Hence, the condition is sufficient as well. //

Theorem 3.2. Given a scalar x and a vector **a**, vide Eq. (3.3), their scalar multiplication is a vector

$$x\,\mathbf{a} = (xa_1, xa_2, xa_3)\,. \tag{3.5}$$

Proof. Multiplying the vector **a**, given by Eqs. (3.4), by the scalar x, we get

$$x\mathbf{a} = x\,(a_1\,\hat{\mathbf{i}} + a_2\,\hat{\mathbf{j}} + a_3\,\hat{\mathbf{k}}) = (xa_1)\,\hat{\mathbf{i}} + (xa_2)\,\hat{\mathbf{j}} + (xa_3)\,\hat{\mathbf{k}} = (xa_1, xa_2, xa_3).\ //$$

Corollary 3.1. The vector $-\,\mathbf{a} = (-\,a_1, -\,a_2, -\,a_3)$ is the negative of the vector **a** given by Eq. (3.3). //

Definition 3.2. The vector sum of two vectors in Eq. (3.3) is defined by

$$\mathbf{a} + \mathbf{b} = (a_1 + b_1, a_2 + b_2, a_3 + b_3), \tag{3.6}$$

whereas the scalar multiple of a vector is defined by Eq. (3.5).

§ 4. Products of Vectors

4.1. Dot product of two vectors

Let **a** and **b** be two non-null vectors inclined at an angle θ (measured from **a** to **b**). Their *dot product* **a . b** is defined by

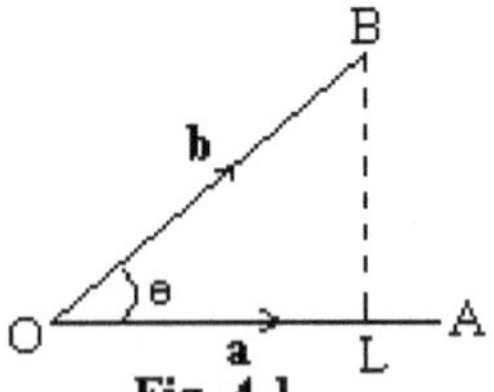

Fig. 4.1

$$\mathbf{a}\,.\,\mathbf{b} = |\,\mathbf{a}\,|\,|\,\mathbf{b}\,|\cos\theta. \tag{4.1}$$

Being a scalar, it is also called the *scalar* (or *inner*) *product* of two vectors. For $\cos(-\,\theta) = \cos\theta$,

$$\mathbf{b}\,.\,\mathbf{a} = |\,\mathbf{b}\,|\,|\,\mathbf{a}\,|\cos(-\,\theta) = \mathbf{a}\,.\,\mathbf{b}\,. \tag{4.2}$$

Thus, the dot product is *commutative*. Replacing **b** by **a** in Eq. (4.1), there follows

$$\mathbf{a}.\,\mathbf{a} = |\,\mathbf{a}\,|\,|\,\mathbf{a}\,|\cos 0 = |\,\mathbf{a}\,|^2 \Rightarrow |\,\mathbf{a}\,| = +\sqrt{(\mathbf{a}\,.\,\mathbf{a})}\,. \tag{4.3}$$

Also, the unit vectors $\hat{\mathbf{i}}$, $\hat{\mathbf{j}}$, $\hat{\mathbf{k}}$ acting along the coordinate axes satisfy

$$\hat{\mathbf{i}}\,.\,\hat{\mathbf{i}} = \hat{\mathbf{j}}\,.\,\hat{\mathbf{j}} = \hat{\mathbf{k}}\,.\,\hat{\mathbf{k}} = 1, \quad \hat{\mathbf{i}}\,.\,\hat{\mathbf{j}} = \hat{\mathbf{j}}\,.\,\hat{\mathbf{i}} = \hat{\mathbf{j}}\,.\,\hat{\mathbf{k}} = \hat{\mathbf{k}}\,.\,\hat{\mathbf{j}} = \hat{\mathbf{k}}\,.\,\hat{\mathbf{i}} = \hat{\mathbf{i}}\,.\,\hat{\mathbf{k}} = 0. \tag{4.4}$$

Accordingly, the dot product of the vector in Eq. (3.1) with itself yields

$$\mathbf{r}\,.\,\mathbf{r} = |\,\mathbf{r}\,|^{2} = (x\,\hat{\mathbf{i}} + y\,\hat{\mathbf{j}} + z\hat{\mathbf{k}})\,.\,(x\,\hat{\mathbf{i}} + y\,\hat{\mathbf{j}} + z\hat{\mathbf{k}}) = x^2 + y^2 + z^2 \qquad (4.5)$$

$\Rightarrow$

$$|\,\mathbf{r}\,| = \sqrt{(x^2 + y^2 + z^2)} = r \text{ (say)}. \qquad (4.6)$$

The unit vector along **r** is

$$\hat{\mathbf{r}} = \mathbf{r}\,/\,|\,\mathbf{r}\,| = (x\,/\,r)\,\hat{\mathbf{i}} + (y\,/\,r)\,\hat{\mathbf{j}} + (z\,/\,r)\,\hat{\mathbf{k}}\,. \qquad (4.7)$$

Theorem 4.1. Two non-null vectors in Eq. (3.3) are orthogonal to each other if

$$\mathbf{a}\,.\,\mathbf{b} = 0. \qquad (4.8)$$

Proof. For mutually perpendicular vectors their dot product given by Eq. (4.1) becomes

$$\mathbf{a}\,.\,\mathbf{b} = |\,\mathbf{a}\,|\,|\,\mathbf{b}\,| \cos 90 = 0.$$

This proves the necessary part. Conversely, when Eq. (4.8) holds there follows $\cos\theta = 0$, as the vectors being non-null. Hence, $\theta = 90^\circ$. //

Theorem 4.2. The dot product of two vectors in Eq. (3.4) is

$$\mathbf{a}\,.\,\mathbf{b} = a_1\,b_1 + a_2\,b_2 + a_3\,b_3\,. \qquad (4.9)$$

Proof. Forming the dot product of two vectors and putting from Eq. (4.4) we get the result. //

Corollary 4.1. The magnitude of the vector **a**, given by Eq. (3.3), is

$$|\,\mathbf{a}\,| = \sqrt{(a_1^2 + a_2^2 + a_3^2)}\,. \qquad (4.10)$$

Proof. Replacing **b** by **a** in Eq. (4.9) and using Eq. (4.3) we obtain the desired result. //

Theorem 4.3. Angle between two vectors **a** and **b** can be measured by

$$\cos\theta = \mathbf{a}\,.\,\mathbf{b}\,/\,|\,\mathbf{a}\,|\,|\,\mathbf{b}\,|$$

$$= (a_1b_1 + a_2b_2 + a_3b_3)\,/\,\sqrt{(a_1^2 + a_2^2 + a_3^2)}\,\sqrt{(b_1^2 + b_2^2 + b_3^2)}\,. \qquad (4.11)$$

Proof. Analogous to Eq. (4.10) we also have

$$|\mathbf{b}| = \sqrt{(b_1^2 + b_2^2 + b_3^2)}\,. \qquad (4.12)$$

Putting from Eqs. (4.9), (4.10) and (4.12) in Eq. (4.1), we get the result. //

Corollary 4.2. The vectors given by Eqs. (3.3) are orthogonal to each other if

$$a_1 b_1 + a_2 b_2 + a_3 b_3 = 0. \qquad (4.13)$$

Proof. The result follows immediately from Eqs. (4.8) and (4.9). //

Definition 4.1. Mutually orthogonal unit vectors are called *orthonormal* vectors.

The unit vectors $\hat{\mathbf{i}}, \hat{\mathbf{j}}, \hat{\mathbf{k}}$ acting along the respective rectangular coordinate axes Ox, Oy and Oz are orthonormal vectors.

Definition 4.2. Let BL be drawn perpendicular from B to $\overline{\text{OA}} = \mathbf{a}$ in Fig. 4.1. The length OL given by

$$\text{OL} = (\text{OB}) \cos\theta = |\mathbf{b}| \cos\theta = |\mathbf{b}|\,|\mathbf{a}|\,(\cos\theta)\,/\,|\mathbf{a}|$$

$$= \mathbf{b}\,.\,\mathbf{a}\,/\,a = \overline{\text{OB}} \cdot \overline{\text{OA}}\,/\,\text{OA} \qquad (4.14)$$

is called the *resolved part of vector* **b** *along the vector* **a**.

Theorem 4.4. The dot product is distributive over vector addition:

$$\mathbf{a}\,.\,(\mathbf{b} + \mathbf{c}) = \mathbf{a}\,.\,\mathbf{b} + \mathbf{a}\,.\,\mathbf{c}. \qquad (4.15)$$

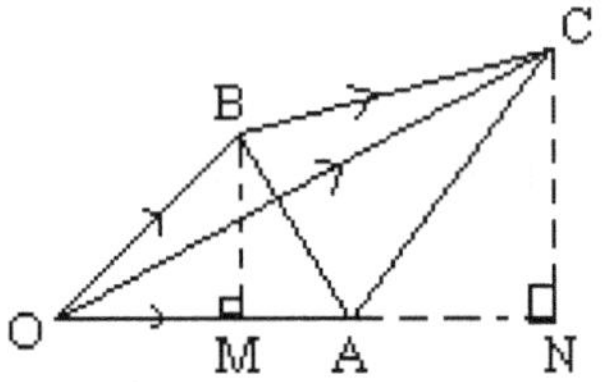

Fig. 4.2

Proof. Let **a**, **b**, **c** be the position vectors of three points A, B, C with respect to some origin O. Join OC and draw the perpendiculars BM and CN to $\overline{\text{OA}} = \mathbf{a}$, so that OM, MN and ON are the resolved parts of vectors $\overline{\text{OB}} = \mathbf{b}$, $\overline{\text{BC}} = \mathbf{c}$ and $\overline{\text{OC}} = \overline{\text{OB}} + \overline{\text{BC}} = \mathbf{b} + \mathbf{c}$. Therefore, the LHS of Eq. (4.15) is

$$a\,\{\text{resolved part of } (\mathbf{b} + \mathbf{c}) \text{ along } \mathbf{a}\} = a\,(\text{ON})$$

$$= a\,(\text{OM} + \text{MN}) = a\,(\text{OM}) + a\,(\text{MN})$$

$$= a\,(\text{resolved part of } \mathbf{b} \text{ along } \mathbf{a}) + a\,(\text{resolved part of } \mathbf{c} \text{ along } \mathbf{a})$$

$= \mathbf{a} \cdot \mathbf{b} + \mathbf{a} \cdot \mathbf{c} =$ RHS of Eq. (4.15).

4.2. Cross product of two vectors

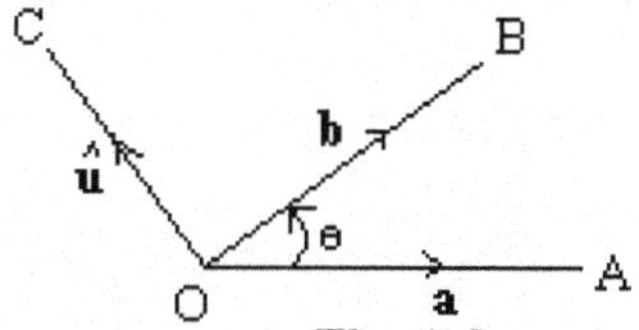

Fig. 4.3

The product of two vectors **a** and **b** defined by

$$\mathbf{a} \times \mathbf{b} \equiv |\mathbf{a}||\mathbf{b}|(\sin\theta)\,\hat{\mathbf{u}}, \qquad (4.16)$$

where $\hat{\mathbf{u}}$ is the unit vector orthogonal to both **a** and **b** such that **a**, **b** and $\hat{\mathbf{u}}$ form a right-handed system. The cross product being a vector is also called the *vector* (or *outer*) *product*. In contrary to Eq. (4.2), the cross product of two vectors is *anti-commutative* (i.e. skew-symmetric):

$$\mathbf{b} \times \mathbf{a} = |\mathbf{b}||\mathbf{a}|(\sin\theta)(-\hat{\mathbf{u}}) = -|\mathbf{b}||\mathbf{a}|(\sin\theta)\,\hat{\mathbf{u}} = -(\mathbf{a} \times \mathbf{b}). \quad (4.17)$$

Analogous to the relations (4.3) and (4.4) there also hold:

$$\left.\begin{array}{l} \mathbf{a} \times \mathbf{a} = |\mathbf{a}|^2(\sin 0)\,\hat{\mathbf{u}} = \mathbf{0}, \quad \hat{\mathbf{i}} \times \hat{\mathbf{i}} = \hat{\mathbf{j}} \times \hat{\mathbf{j}} = \hat{\mathbf{k}} \times \hat{\mathbf{k}} = \mathbf{0}, \\ \hat{\mathbf{i}} \times \hat{\mathbf{j}} = -\hat{\mathbf{j}} \times \hat{\mathbf{i}} = \hat{\mathbf{k}}, \hat{\mathbf{j}} \times \hat{\mathbf{k}} = -\hat{\mathbf{k}} \times \hat{\mathbf{j}} = \hat{\mathbf{i}}, \ \hat{\mathbf{k}} \times \hat{\mathbf{i}} = -\hat{\mathbf{i}} \times \hat{\mathbf{k}} = \hat{\mathbf{j}}. \end{array}\right\} \quad (4.18)$$

Note 4.1. The unit vectors $\hat{\mathbf{i}}, \hat{\mathbf{j}}, \hat{\mathbf{k}}$ acting along the respective coordinate axes form a right-handed system, and so do the triads: $\hat{\mathbf{j}}, \hat{\mathbf{k}}, \hat{\mathbf{i}}$ and $\hat{\mathbf{k}}, \hat{\mathbf{i}}, \hat{\mathbf{j}}$. An interchange of any two vectors in these triads makes them left-handed. Thus, $\hat{\mathbf{j}}, \hat{\mathbf{i}}, \hat{\mathbf{k}}$; $\hat{\mathbf{k}}, \hat{\mathbf{j}}, \hat{\mathbf{i}}$; and $\hat{\mathbf{i}}, \hat{\mathbf{k}}, \hat{\mathbf{j}}$ are in the left-handed system. On the other hand, $\hat{\mathbf{j}}, \hat{\mathbf{i}}, -\hat{\mathbf{k}}$; $\hat{\mathbf{k}}, \hat{\mathbf{j}}, -\hat{\mathbf{i}}$; and $\hat{\mathbf{i}}, \hat{\mathbf{k}}, -\hat{\mathbf{j}}$ again form right-handed systems.

Theorem 4.5. The cross product of two vectors **a** and **b**, given by Eqs. (3.3), is

$$\mathbf{a} \times \mathbf{b} = \begin{vmatrix} \hat{\mathbf{i}} & \hat{\mathbf{j}} & \hat{\mathbf{k}} \\ a_1 & a_2 & a_3 \\ b_1 & b_2 & b_3 \end{vmatrix}$$

$$= (a_2 b_3 - a_3 b_2)\,\hat{\mathbf{i}} + (a_3 b_1 - a_1 b_3)\,\hat{\mathbf{j}} + (a_1 b_2 - a_2 b_1)\,\hat{\mathbf{k}}. \qquad (4.19)$$

Proof. $\mathbf{a} \times \mathbf{b} = (a_1\hat{\mathbf{i}} + a_2\hat{\mathbf{j}} + a_3\hat{\mathbf{k}}) \times (b_1\hat{\mathbf{i}} + b_2\hat{\mathbf{j}} + b_3\hat{\mathbf{k}})$

$$= (a_1 b_2 - a_2 b_1)(\hat{\mathbf{i}} \times \hat{\mathbf{j}}) + (a_2 b_3 - a_3 b_2)(\hat{\mathbf{j}} \times \hat{\mathbf{k}}) + (a_3 b_1 - a_1 b_3)(\hat{\mathbf{k}} \times \hat{\mathbf{i}})$$

$$= (a_1 b_2 - a_2 b_1)\, \hat{\mathbf{k}} + (a_2 b_3 - a_3 b_2)\, \hat{\mathbf{i}} + (a_3 b_1 - a_1 b_3)\, \hat{\mathbf{j}},$$

by Eq. (4.18). //

Theorem 4.6. Angle between two vectors can also be measured by

$$\sin^2\theta = \{(a_1 b_2 - a_2 b_1)^2 + (a_2 b_3 - a_3 b_2)^2 + (a_3 b_1 - a_1 b_3)^2\}$$

$$\div\ (a_1^2 + a_2^2 + a_3^2)\,(b_1^2 + b_2^2 + b_3^2). \tag{4.20}$$

Proof. Forming the dot product of Eq. (4.19) with itself, using Eq. (4.16) and putting from Eq. (4.4) we get

$$|\,\mathbf{a}\,|^2\,|\,\mathbf{b}\,|^2\,(\sin^2\theta)\,(\hat{\mathbf{u}}\,.\,\hat{\mathbf{u}}) = \sum (a_1 b_2 - a_2 b_1)^2.$$

Further, putting from Eqs. (4.10), (4.12) and $\hat{\mathbf{u}}\,.\,\hat{\mathbf{u}} = 1$, we get the result. //

Corollary 4.3. The vectors, given by Eq. (3.3), are *parallel* if

$$\mathbf{a} \times \mathbf{b} = \mathbf{0} \ \Leftrightarrow\ a_1/b_1 = a_2/b_2 = a_3/b_3. \tag{4.21}$$

Proof. It follows from Eq. (4.16) that two non-null vectors **a** and **b** are parallel if the angle between them is zero. Hence, the first condition follows from Eq. (4.16) itself. Further, it follows from Eq. (4.20) that $\theta = 0$ if the numerator in the RHS of Eq. (4.20) vanishes identically, which is so if each individual term vanishes therein giving rise to the second condition. //

Theorem 4.7. For any scalar x and two vectors **a** and **b**, there hold

$$x\,(\mathbf{a} \times \mathbf{b}) = (x\,\mathbf{a}) \times \mathbf{b} = (\mathbf{a} \times x\,\mathbf{b}). \tag{4.22}$$

Proof being simple is left to the reader.

Theorem 4.8. For any two vectors **a** and **b** there hold

$$\mathbf{a}\,.\,(\mathbf{a} \times \mathbf{b}) = 0 = \mathbf{b}\,.\,(\mathbf{a} \times \mathbf{b}). \tag{4.23}$$

Proof. By definition, **a** × **b** is orthogonal to both **a** and **b**. Hence, the results follow from Theorem 4.1. //

Theorem 4.9. The operation of cross product is distributive over vector addition:

$$\mathbf{a} \times (\mathbf{b} + \mathbf{c}) = \mathbf{a} \times \mathbf{b} + \mathbf{a} \times \mathbf{c}. \tag{4.24}$$

Proof. Computing the cross product of $\mathbf{a} = (a_1, a_2, a_3)$ and

$$\mathbf{b} + \mathbf{c} = (b_1 + c_1, b_2 + c_2, b_3 + c_3), \tag{4.25}$$

by Eq. (4.19), we evaluate the LHS of Eq. (4.24):

$$\text{LHS} = \{a_2 (b_3 + c_3) - a_3 (b_2 + c_2)\}\hat{\mathbf{i}} + \{a_3 (b_1 + c_1) - a_1 (b_3 + c_3)\}\, \hat{\mathbf{j}}$$

$$+ \{a_1 (b_2 + c_2) - a_2 (b_1 + c_1)\}\, \hat{\mathbf{k}}\,.$$

Similarly, computing the cross product $\mathbf{a} \times \mathbf{c}$ and adding it to that given by Eq. (4.19), the sum may be seen same as above. //

Example 4.1. Let **a** and **b** be two vectors with magnitudes a and b respectively. Show that

$$(\mathbf{a} + \mathbf{b}) \,.\, (\mathbf{a} - \mathbf{b}) = a^2 - b^2. \tag{4.26}$$

Solution. Applying distributive law vide Eq.(4.15) and the commutative property of the *dot product* we prove the result. //

Example 4.2. Let **a**, **b**, **c** be three mutually perpendicular vectors of same magnitude. Show that their sum vector $\mathbf{a} + \mathbf{b} + \mathbf{c}$ is equally inclined to each of the vectors **a**, **b**, **c**.

Solution. The magnitude of the vector $\mathbf{a} + \mathbf{b} + \mathbf{c}$, by Eq. (4.3), is

$$|\,\mathbf{a} + \mathbf{b} + \mathbf{c}\,| = \sqrt{\{(\mathbf{a} + \mathbf{b} + \mathbf{c}) \,.\, (\mathbf{a} + \mathbf{b} + \mathbf{c})\}} = \sqrt{(a^2 + b^2 + c^2)} = a\sqrt{3},$$

where the relations analogous to those given by Eqs. (4.4) satisfied by mutually orthogonal vectors **a**, **b**, **c** are used. Therefore, we get

$$(\mathbf{a} + \mathbf{b} + \mathbf{c}) \,.\, \mathbf{a} = a^2 = (a^2 \sqrt{3}) \cos \theta_1 \quad \Rightarrow \quad \cos \theta_1 = 1/\sqrt{3}.$$

Similarly, we can show that $\cos \theta_2 = \cos \theta_3 = 1/\sqrt{3}$. //

Example 4.3. For any three vectors **a**, **b**, **c** prove that

$$\mathbf{a} \times (\mathbf{b} + \mathbf{c}) + \mathbf{b} \times (\mathbf{c} + \mathbf{a}) + \mathbf{c} \times (\mathbf{a} + \mathbf{b}) = \mathbf{0}. \tag{4.27}$$

Solution. Applying Eq. (4.24), the LHS of above identity becomes

$$\mathbf{a} \times \mathbf{b} + \mathbf{a} \times \mathbf{c} + \mathbf{b} \times \mathbf{c} + \mathbf{b} \times \mathbf{a} + \mathbf{c} \times \mathbf{a} + \mathbf{c} \times \mathbf{b},$$

which, for skew-symmetric property of cross products, vanishes. //

Example 4.4. For any vectors **a** and **b** of magnitudes a and b, prove

$$(\mathbf{a} \times \mathbf{b})^2 + (\mathbf{a} \,.\, \mathbf{b})^2 = (a\,b)^2. \tag{4.28}$$

Solution. Squaring the results in Eqs. (4.1) and (4.16) and adding them, the LHS of Eq. (4.28) becomes

$$|\,\mathbf{a}\,|^2\,|\,\mathbf{b}\,|^2 (\sin^2\theta + \cos^2\theta) = a^2\,b^2. \;//$$

Example 4.5. For any two vectors **a** and **b**, show that

$$(\mathbf{a} - \mathbf{b}) \times (\mathbf{a} + \mathbf{b}) = 2\,(\mathbf{a} \times \mathbf{b}). \tag{4.29}$$

Solution. Applying the distributive law given by Eq. (4.24), the LHS of above identity simplifies to

$$\mathbf{a} \times (\mathbf{a} + \mathbf{b}) - \mathbf{b} \times (\mathbf{a} + \mathbf{b}) = \mathbf{a} \times \mathbf{b} - \mathbf{b} \times \mathbf{a} = 2\,(\mathbf{a} \times \mathbf{b}). \;//$$

§ 5. Products of three vectors

As seen in the previous section, the cross product of two vectors **a** and **b** is a vector. So, its further products with a third vector, say

$$\mathbf{c} = (c_1, c_2, c_3) \tag{5.1}$$

can also be defined in the following two ways:

(i) dot product: $(\mathbf{a} \times \mathbf{b}).\,\mathbf{c}$, denoted by $[\mathbf{a}\ \ \mathbf{b}\ \ \mathbf{c}]$. Being a scalar, it is also called a *scalar* (*triple*) *product* of vectors **a**, **b**, **c**.

(ii) cross product: $$(\mathbf{a} \times \mathbf{b}) \times \mathbf{c} \equiv \mathbf{d} \text{ (say)} \tag{5.2}$$

is defined in analogy with Eq. (4.16). Being a vector, it is also called a *vector* (*triple*) *product* of vectors **a**, **b**, **c**.

(iii) As seen in § 4, the dot product of any two of the vectors **a**, **b**, **c**, say **a . b** is scalar. Hence, its product with the third vector **c** is just a scalar multiplication (**a . b**) **c**, which is already discussed in § 2 above.

5.1. Properties of [a b c]

Theorem 5.1. The dot product of three vectors **a**, **b** and **c** given by Eqs. (3.3) and (5.1) is

$$[\mathbf{a}\ \mathbf{b}\ \mathbf{c}] = \begin{vmatrix} a_1 & a_2 & a_3 \\ b_1 & b_2 & b_3 \\ c_1 & c_2 & c_3 \end{vmatrix}. \tag{5.3}$$

Proof. Forming dot product of the vectors given by Eqs. (4.19) and (5.1), we get

$$[\mathbf{a}\ \mathbf{b}\ \mathbf{c}] = (a_2 b_3 - a_3 b_2)\, c_1 + (a_3 b_1 - a_1 b_3)\, c_2 + (a_1 b_2 - a_2 b_1)\, c_3,$$

which is just the expansion of the determinant in Eq. (5.3). //

Corollary 5.1. The scalar triple product of orthonormal vectors $\hat{\mathbf{i}}$, $\hat{\mathbf{j}}$, $\hat{\mathbf{k}}$ has value 1:

$$[\hat{\mathbf{i}}\ \hat{\mathbf{j}}\ \hat{\mathbf{k}}] = 1. \tag{5.4}$$

Proof. $[\hat{\mathbf{i}}\ \hat{\mathbf{j}}\ \hat{\mathbf{k}}] = (\hat{\mathbf{i}} \times \hat{\mathbf{j}}) \cdot \hat{\mathbf{k}} = \hat{\mathbf{k}} \cdot \hat{\mathbf{k}} = 1$, by Eqs. (4.4) and (4.18). //

Theorem 5.2. The scalar triple product [**a b c**] gives the volume V of a cuboid having **a**, **b**, **c** as the coterminus edges; and it satisfies

$$V = [\mathbf{a}\ \mathbf{b}\ \mathbf{c}] = [\mathbf{b}\ \mathbf{c}\ \mathbf{a}] = [\mathbf{c}\ \mathbf{a}\ \mathbf{b}]. \tag{5.5}$$

Corollary 5.2. Three vectors are coplanar (i.e. lying in the same plane) iff their scalar triple product is zero:

$$[\mathbf{a}\ \mathbf{b}\ \mathbf{c}] = 0. \tag{5.6}$$

Proof. The result follows immediately from Eq. (5.5) as the parallel-opiped formed by three coplanar vectors is of zero volume.

Alternately, if **c** lies in the plane containing **a** and **b** so it is orthogonal to $\mathbf{a} \times \mathbf{b}$ as the latter vector is orthogonal to that plane. This makes $(\mathbf{a} \times \mathbf{b}) \, . \, \mathbf{c} = 0$. //

Corollary 5.3. The positions of *dot* and *cross* operators are interchangeable in the product $(\mathbf{a} \times \mathbf{b}) \, . \, \mathbf{c}$:

$$(\mathbf{a} \times \mathbf{b}) \, . \, \mathbf{c} \;=\; \mathbf{a} \, . \, (\mathbf{b} \times \mathbf{c}) \;=\; [\mathbf{a} \; \mathbf{b} \; \mathbf{c}]. \tag{5.7}$$

Proof. The proof follows from the commutative property of the dot product of two vectors and the Eqs. (5.5). //

For anti-symmetric property of the cross product $\mathbf{a} \times \mathbf{b}$ exhibited by Eq. (4.17), Eqs. (5.5) and (5.7) also yield

$$[\mathbf{a} \;\; \mathbf{b} \;\; \mathbf{c}] = -\,[\,\mathbf{b} \; \mathbf{a} \; \mathbf{c}] \;=\; -\,[\mathbf{c} \;\; \mathbf{b} \;\; \mathbf{a}] \;=\; -\,[\mathbf{a} \;\; \mathbf{c} \;\; \mathbf{b}]. \tag{5.8}$$

Corollary 5.4. If any two vectors in $[\mathbf{a} \;\; \mathbf{b} \;\; \mathbf{c}]$ are equal (or collinear) the scalar triple product vanishes:

$$[\mathbf{a} \;\; \mathbf{b} \;\; \mathbf{b}] = [\mathbf{a} \;\; \mathbf{a} \;\; \mathbf{c}] = [\mathbf{a} \;\; \mathbf{b} \;\; \mathbf{a}] = 0. \tag{5.9}$$

Proof. $\mathbf{a} \times \mathbf{b}$ being perpendicular to **b** (as well as to **a**) its dot products with both **b** and **a** vanish:

$$(\mathbf{a} \times \mathbf{b}) \, . \, \mathbf{b} = \;[\mathbf{a} \;\; \mathbf{b} \;\; \mathbf{b}] = \;0, \qquad (\mathbf{a} \times \mathbf{b}) \, . \, \mathbf{a} = \;[\mathbf{a} \;\; \mathbf{b} \;\; \mathbf{a}] = \;0.$$

Similarly, $\mathbf{a} \times \mathbf{c}$ being perpendicular to **a** its dot product with **a** vanishes:

$$(\mathbf{a} \times \mathbf{c}) \, . \, \mathbf{a} = \mathbf{a} \, . \, (\mathbf{a} \times \mathbf{c}) \;=\; [\mathbf{a} \;\; \mathbf{a} \;\; \mathbf{c}] = \;0. \; //$$

5.2. Properties of $(\mathbf{a} \times \mathbf{b}) \times \mathbf{c}$:

As seen in § 4, the vector $\mathbf{a} \times \mathbf{b}$ is orthogonal to both **a** and **b**. So, it is normal to the plane containing **a** and **b**. Analogously, the vector **d** given by Eq. (5.2) is orthogonal to both $\mathbf{a} \times \mathbf{b}$ and **c**. Hence, **d** lies in the plane containing **a** and **b**. Therefore, it can be linearly expressed in terms of **a** and **b**:

$$\mathbf{d} \;=\; l\mathbf{a} + m\mathbf{b}, \tag{5.10}$$

where l, m are some scalars to be determined. Forming the dot product of this vector with **c** and noting that $\mathbf{d} \, . \, \mathbf{c} = 0$, we derive

$$l\,(\mathbf{a}\,.\,\mathbf{c}) + m\,(\mathbf{b}\,.\,\mathbf{c}) = 0 \quad \Rightarrow \quad l\,/\,(\mathbf{b}\,.\,\mathbf{c}) = -\,m\,/\,(\mathbf{a}\,.\,\mathbf{c}) = n \text{ (say)}$$

$\Rightarrow$

$$l = n\,(\mathbf{b}\,.\,\mathbf{c}) \quad \text{and} \quad m = -\,n\,(\mathbf{a}\,.\,\mathbf{c}).$$

Accordingly, Eq. (5.10) reduces to

$$(\mathbf{a} \times \mathbf{b}) \times \mathbf{c} = n\,(\mathbf{b}\,.\,\mathbf{c})\,\mathbf{a} - n\,(\mathbf{a}\,.\,\mathbf{c})\,\mathbf{b}. \tag{5.11}$$

Particularly, choosing $\mathbf{a} = \hat{\mathbf{i}}$ and $\mathbf{b} = \mathbf{c} = \hat{\mathbf{j}}$, where $\hat{\mathbf{i}}$, $\hat{\mathbf{j}}$ are mutually orthogonal unit vectors, Eq. (5.9) reduces to

$$(\hat{\mathbf{i}} \times \hat{\mathbf{j}}) \times \hat{\mathbf{j}} = \hat{\mathbf{k}} \times \hat{\mathbf{j}} = -\,\hat{\mathbf{i}},$$

or, by Eq. (5.11)

$$= n\,(\hat{\mathbf{j}}\,.\,\hat{\mathbf{j}})\,\hat{\mathbf{i}} - n\,(\hat{\mathbf{i}}\,.\,\hat{\mathbf{j}})\,\hat{\mathbf{j}} = n\,\hat{\mathbf{i}} \quad \Rightarrow \quad n = -\,1,$$

where Eqs. (4.4) and (4.18) are used. Hence, Eq. (5.11) finally assumes the form

$$(\mathbf{a} \times \mathbf{b}) \times \mathbf{c} = (\mathbf{a}\,.\,\mathbf{c})\,\mathbf{b} - (\mathbf{b}\,.\,\mathbf{c})\,\mathbf{a}. \tag{5.12}$$

§ 6. Products of four vectors

Given four vectors **a**, **b**, **c**, **d** their cross products $\mathbf{a} \times \mathbf{b}$ and $\mathbf{c} \times \mathbf{d}$ also define two more products:

(i) Dot or scalar product $(\mathbf{a} \times \mathbf{b})\,.\,(\mathbf{c} \times \mathbf{d})$, and

(ii) Cross or vector product $(\mathbf{a} \times \mathbf{b}) \times (\mathbf{c} \times \mathbf{d})$.

Theorem 6.1. We have

$$(\mathbf{a} \times \mathbf{b})\,.\,(\mathbf{c} \times \mathbf{d}) = \begin{vmatrix} \mathbf{a}.\,\mathbf{c} & \mathbf{a}.\,\mathbf{d} \\ \mathbf{b}.\,\mathbf{c} & \mathbf{b}.\,\mathbf{d} \end{vmatrix} = (\mathbf{a}.\,\mathbf{c})\,(\mathbf{b}.\,\mathbf{d}) - (\mathbf{a}.\,\mathbf{d})\,(\mathbf{b}.\,\mathbf{c}). \tag{6.1}$$

Proof. Setting

$$\mathbf{a} \times \mathbf{b} = \mathbf{p}, \tag{6.2}$$

the product in LHS of Eq. (6.1) is

$$\mathbf{p}\,.\,(\mathbf{c} \times \mathbf{d}) = (\mathbf{p} \times \mathbf{c})\,.\,\mathbf{d}, \qquad \text{by Cor. (5.3)}$$

$$= \{\,(\mathbf{a} \times \mathbf{b}) \times \mathbf{c}\,\}\,.\,\mathbf{d} = \{\,(\mathbf{a}.\,\mathbf{c})\,\mathbf{b} - (\mathbf{b}.\,\mathbf{c})\,\mathbf{a}\,\}.\,\mathbf{d}, \qquad \text{by Eq. (5.12)}$$

which is same as on the RHS of Eq. (6.1). //

Theorem 6.2. We have

$$(\mathbf{a} \times \mathbf{b}) \times (\mathbf{c} \times \mathbf{d}) = [\mathbf{a}\ \mathbf{b}\ \mathbf{d}]\ \mathbf{c} - [\mathbf{a}\ \mathbf{b}\ \mathbf{c}]\ \mathbf{d} = [\mathbf{a}\ \mathbf{c}\ \mathbf{d}]\ \mathbf{b} - [\mathbf{b}\ \mathbf{c}\ \mathbf{d}]\ \mathbf{a}. \quad (6.3)$$

Proof. (i) The vector on LHS of Eq. (6.3), for Eq. (6.2), reduces to

$$\mathbf{p} \times (\mathbf{c} \times \mathbf{d}) = (\mathbf{p} \cdot \mathbf{d})\ \mathbf{c} - (\mathbf{p} \cdot \mathbf{c})\ \mathbf{d}, \qquad \text{by Eq. (5.12)}$$

$$= [\mathbf{a}\ \mathbf{b}\ \mathbf{d}]\ \mathbf{c} - [\mathbf{a}\ \mathbf{b}\ \mathbf{c}]\ \mathbf{d}, \qquad \text{for Eq. (6.2).}$$

(ii) On the other hand, putting

$$\mathbf{c} \times \mathbf{d} = \mathbf{q}, \quad (6.4)$$

the product in LHS of Eq. (6.3) reduces to

$$(\mathbf{a} \times \mathbf{b}) \times \mathbf{q} = (\mathbf{a} \cdot \mathbf{q})\ \mathbf{b} - (\mathbf{b} \cdot \mathbf{q})\ \mathbf{a}, \qquad \text{again by Eq.(5.12)}$$

$$= [\mathbf{a}\ \mathbf{c}\ \mathbf{d}]\ \mathbf{b} - [\mathbf{b}\ \mathbf{c}\ \mathbf{d}]\ \mathbf{a}, \qquad \text{for Eq. (6.4)}$$

establishing the alternate value in Eq. (6.3). //

Corollary 6.1. Any vector **d** can be linearly expressed in terms of three non-coplanar (i.e. linearly independent) vectors **a, b, c**:

$$\mathbf{d} = \{ [\mathbf{b}\ \mathbf{c}\ \mathbf{d}]\ \mathbf{a} + [\mathbf{c}\ \mathbf{a}\ \mathbf{d}]\ \mathbf{b} + [\mathbf{a}\ \mathbf{b}\ \mathbf{d}]\ \mathbf{c} \} / [\mathbf{a}\ \mathbf{b}\ \mathbf{c}]. \quad (6.5)$$

Proof. Alternate values for the vector product in previous theorem determine the fourth vector **d** as a linear combination of the remaining vectors:

$$-[\mathbf{a}\ \mathbf{b}\ \mathbf{c}]\ \mathbf{d} = [\mathbf{a}\ \mathbf{c}\ \mathbf{d}]\ \mathbf{b} - [\mathbf{b}\ \mathbf{c}\ \mathbf{d}]\ \mathbf{a} - [\mathbf{a}\ \mathbf{b}\ \mathbf{d}]\ \mathbf{c}.$$

Dividing it by – [**a b c**], and noting Eq. (5.8), above equation assumes the desired form. //

The relation (6.5) justifies the linear dependence of all four vectors **a, b, c, d**, and we have the:

Note 6.1. Any four vectors in a 3-dimensional space are always linearly dependent.

Example 6.1. Let $\hat{\mathbf{a}}, \hat{\mathbf{b}}, \hat{\mathbf{c}}, \hat{\mathbf{d}}$ be the unit vectors acting along the forces (of magnitudes) F_1, F_2, F_3, F_4 which are in equilibrium. Prove the relations

$$F_1 / [\hat{\mathbf{b}}\ \hat{\mathbf{c}}\ \hat{\mathbf{d}}] = F_2 / [\hat{\mathbf{c}}\ \hat{\mathbf{a}}\ \hat{\mathbf{d}}] = F_3 / [\hat{\mathbf{a}}\ \hat{\mathbf{b}}\ \hat{\mathbf{d}}] = F_4 / [\hat{\mathbf{b}}\ \hat{\mathbf{a}}\ \hat{\mathbf{c}}]. \quad (6.6)$$

Solution. Forces being in equilibrium their resultant force is zero:

$$F_1 \hat{\mathbf{a}} + F_2 \hat{\mathbf{b}} + F_3 \hat{\mathbf{c}} + F_4 \hat{\mathbf{d}} = \mathbf{0}. \quad (6.7)$$

Forming its dot product with the vector $\hat{\mathbf{b}} \times \hat{\mathbf{c}}$, and putting from Eqs. (5.9), it yields

$$F_1 [\hat{\mathbf{a}}\ \hat{\mathbf{b}}\ \hat{\mathbf{c}}] + F_4 [\hat{\mathbf{d}}\ \hat{\mathbf{b}}\ \hat{\mathbf{c}}] = \mathbf{0} \Rightarrow F_1 / [\hat{\mathbf{d}}\ \hat{\mathbf{b}}\ \hat{\mathbf{c}}] = -F_4 / [\hat{\mathbf{a}}\ \hat{\mathbf{b}}\ \hat{\mathbf{c}}],$$

which, for Eqs. (5.5) and (5.8) establish the equivalence of the first and last fractions in Eq. (6.6).

Similarly, forming dot products of Eq. (6.7) with vectors $\hat{\mathbf{b}} \times \hat{\mathbf{d}}, \hat{\mathbf{c}} \times \hat{\mathbf{d}}$ and using Eqs. (5.5) and (5.8), we also prove equivalence of the remaining fractions in Eq. (6.6). //

§ 7. Reciprocal systems of vectors

With three non-coplanar vectors **a, b, c** we now define three new vectors:

$$\left.\begin{aligned} {}^{-1}\mathbf{a} = (\mathbf{b} \times \mathbf{c}) / [\mathbf{a}\ \mathbf{b}\ \mathbf{c}], \quad {}^{-1}\mathbf{b} = (\mathbf{c} \times \mathbf{a}) / [\mathbf{a}\ \mathbf{b}\ \mathbf{c}], \\ {}^{-1}\mathbf{c} = (\mathbf{a} \times \mathbf{b}) / [\mathbf{a}\ \mathbf{b}\ \mathbf{c}]. \end{aligned}\right\} \quad (7.1)$$

Definition 7.1. The new system of vectors defined above is called reciprocal system of vectors **a, b, c**.

Theorem 7.1. The two systems of vectors satisfy the relations

$$\mathbf{a} \,.\, {}^{-1}\mathbf{a} = \mathbf{b} \,.\, {}^{-1}\mathbf{b} = \mathbf{c} \,.\, {}^{-1}\mathbf{c} = 1, \quad (7.2)$$

and

$$\mathbf{a} \,.\, {}^{-1}\mathbf{b} = \mathbf{a} \,.\, {}^{-1}\mathbf{c} = \mathbf{b} \,.\, {}^{-1}\mathbf{a} = \mathbf{b} \,.\, {}^{-1}\mathbf{c} = \mathbf{c} \,.\, {}^{-1}\mathbf{a} = \mathbf{c} \,.\, {}^{-1}\mathbf{b} = 0. \quad (7.3)$$

Proof. Forming dot products of vectors in Eq. (7.1) with **a, b, c** respectively and noting Eqs. (5.7) and (5.9), the results can be easily derived. //

Analogous to vectors in Eqs. (7.1), let us also define

$$\left.\begin{aligned} \mathbf{a} &= ({}^{-1}\mathbf{b} \times {}^{-1}\mathbf{c}) / [{}^{-1}\mathbf{a}\ {}^{-1}\mathbf{b}\ {}^{-1}\mathbf{c}], \\ \mathbf{b} &= ({}^{-1}\mathbf{c} \times {}^{-1}\mathbf{a}) / [{}^{-1}\mathbf{a}\ {}^{-1}\mathbf{b}\ {}^{-1}\mathbf{c}], \\ \mathbf{c} &= ({}^{-1}\mathbf{a} \times {}^{-1}\mathbf{b}) / [{}^{-1}\mathbf{a}\ {}^{-1}\mathbf{b}\ {}^{-1}\mathbf{c}]. \end{aligned}\right\} \tag{7.4}$$

Forming dot products of these vectors with ${}^{-1}\mathbf{a}$, ${}^{-1}\mathbf{b}$, ${}^{-1}\mathbf{c}$ and using Eqs. (5.7) and (5.9), the results in Eqs. (7.2) and (7.3) may be again established. Thus, analogous to Defn. 7.1, we also have the:

Definition 7.2. The system of vectors vide Eqs. (7.4) is reciprocal to the system of vectors in Eqs. (7.1). Hence, two systems of vectors **a**, **b**, **c** and ${}^{-1}\mathbf{a}$, ${}^{-1}\mathbf{b}$, ${}^{-1}\mathbf{c}$ are reciprocal to each other.

Theorem 7.2. Two reciprocal systems of vectors satisfy the relations

$$[\mathbf{a}\ \mathbf{b}\ \mathbf{c}]\,[{}^{-1}\mathbf{a}\ {}^{-1}\mathbf{b}\ {}^{-1}\mathbf{c}] = 1, \tag{7.5}$$

Proof. Forming cross product of vectors in Eq. (7.1) and putting from Eqs. (6.3) and (5.9), there follows:

$$\begin{aligned} {}^{-1}\mathbf{b} \times {}^{-1}\mathbf{c} &= (\mathbf{c} \times \mathbf{a}) \times (\mathbf{a} \times \mathbf{b}) / [\mathbf{a}\ \mathbf{b}\ \mathbf{c}]^2, \\ &= [\mathbf{c}\ \mathbf{a}\ \mathbf{b}]\,\mathbf{a} / [\mathbf{a}\ \mathbf{b}\ \mathbf{c}]^2 = \mathbf{a} / [\mathbf{a}\ \mathbf{b}\ \mathbf{c}], \qquad \text{by Eq. (5.5).} \end{aligned}$$

Taking its dot product with ${}^{-1}\mathbf{a}$, there results

$$[{}^{-1}\mathbf{a}\ {}^{-1}\mathbf{b}\ {}^{-1}\mathbf{c}] = {}^{-1}\mathbf{a} \,.\, \mathbf{a} / [\mathbf{a}\ \mathbf{b}\ \mathbf{c}] = 1 / [\mathbf{a}\ \mathbf{b}\ \mathbf{c}], \text{ by Eq. (7.2);}$$

which, on multiplication by $[\mathbf{a}\ \mathbf{b}\ \mathbf{c}]$ yields the desired result. //

Evidently, from Eq. (7.5) there also follows the:

Corollary 7.1. If $[\mathbf{a}\ \mathbf{b}\ \mathbf{c}] \neq 0$, so is $[{}^{-1}\mathbf{a}\ {}^{-1}\mathbf{b}\ {}^{-1}\mathbf{c}]$.

Theorem 7.3. The system of orthonormal vectors $\hat{\mathbf{i}}, \hat{\mathbf{j}}, \hat{\mathbf{k}}$ is self-reciprocal system of vectors.

Proof. Constructing the reciprocal system of vectors $\hat{\mathbf{i}}, \hat{\mathbf{j}}, \hat{\mathbf{k}}$, in analogy with Eq. (7.1), we have

$$\left.\begin{aligned} {}^{-1}\hat{\mathbf{i}} &= (\hat{\mathbf{j}} \times \hat{\mathbf{k}})/[\hat{\mathbf{i}}\ \hat{\mathbf{j}}\ \hat{\mathbf{k}}] = \hat{\mathbf{i}}, \\ {}^{-1}\hat{\mathbf{j}} &= (\hat{\mathbf{k}} \times \hat{\mathbf{i}})/[\hat{\mathbf{i}}\ \hat{\mathbf{j}}\ \hat{\mathbf{k}}] = \hat{\mathbf{j}}, \\ {}^{-1}\hat{\mathbf{k}} &= (\hat{\mathbf{i}} \times \hat{\mathbf{j}})/[\hat{\mathbf{i}}\ \hat{\mathbf{j}}\ \hat{\mathbf{k}}] = \hat{\mathbf{k}}, \end{aligned}\right\} \tag{7.6}$$

by Eqs. (4.18) and (5.4). This establishes the result. //

Theorem 7.4. If **a**, **b**, **c** form a basis of the space any vector **d** in the space can be linearly expressed in terms of the base vectors:

$$\mathbf{d} = (\mathbf{d}\,.\,{}^{-1}\mathbf{a})\,\mathbf{a} + (\mathbf{d}\,.\,{}^{-1}\mathbf{b})\,\mathbf{b} + (\mathbf{d}\,.\,{}^{-1}\mathbf{c})\,\mathbf{c}. \tag{7.7}$$

Proof. For reciprocal system of vectors given by Eqs. (7.1), the relation (6.5) can be re-written as above. //

Corollary 7.2. If **a**, **b**, **c** and ${}^{-1}\mathbf{a}$, ${}^{-1}\mathbf{b}$, ${}^{-1}\mathbf{c}$ form two reciprocal systems of vectors any vector **d** in the space can also be written as

$$\mathbf{d} = (\mathbf{d}\,.\,\mathbf{a})\,{}^{-1}\mathbf{a} + (\mathbf{d}\,.\,\mathbf{b})\,{}^{-1}\mathbf{b} + (\mathbf{d}\,.\,\mathbf{c})\,{}^{-1}\mathbf{c}. \tag{7.8}$$

Proof. For Cor. 7.1, the vectors given by Eqs. (7.1), also form a basis of the space and any vector **d** of the space can be expressed in terms of these in analogy with relation (6.5):

$$\mathbf{d} = \{[{}^{-1}\mathbf{b}\ {}^{-1}\mathbf{c}\ \mathbf{d}]\,{}^{-1}\mathbf{a} + [{}^{-1}\mathbf{c}\ {}^{-1}\mathbf{a}\ \mathbf{d}]\,{}^{-1}\mathbf{b} + [{}^{-1}\mathbf{a}\ {}^{-1}\mathbf{b}\ \mathbf{d}]\,{}^{-1}\mathbf{c}\}\,/\,[{}^{-1}\mathbf{a}\ {}^{-1}\mathbf{b}\ {}^{-1}\mathbf{c}].$$

Putting from Eqs. (7.4), the RHS expression of above equation reduces to the form of Eq. (7.8). //

Example 7.1. Find the reciprocal system of vectors **a**, **b** and **a** × **b**.

Solution. Replacing vector **c** by **a** × **b** in Eqs. (7.1), and putting from Eq. (5.12), we get

$$^{-1}\mathbf{a} = \{\mathbf{b} \times (\mathbf{a} \times \mathbf{b})\} / [\mathbf{a}\ \ \mathbf{b}\ \ \mathbf{a} \times \mathbf{b}] = \{(\mathbf{b}.\mathbf{b})\,\mathbf{a} - (\mathbf{b}.\mathbf{a})\,\mathbf{b}\} / [\mathbf{a}\ \ \mathbf{b}\ \ \mathbf{a} \times \mathbf{b}],$$

$$^{-1}\mathbf{b} = \{(\mathbf{a} \times \mathbf{b}) \times \mathbf{a}) / [\mathbf{a}\ \ \mathbf{b}\ \ \mathbf{a} \times \mathbf{b}] = \{(\mathbf{a}.\mathbf{a})\,\mathbf{b} - (\mathbf{b}.\mathbf{a})\,\mathbf{a}\} / [\mathbf{a}\ \ \mathbf{b}\ \ \mathbf{a} \times \mathbf{b}],$$

and

$$^{-1}(\mathbf{a} \times \mathbf{b}) = (\mathbf{a} \times \mathbf{b}) / [\mathbf{a}\ \ \mathbf{b}\ \ \mathbf{a} \times \mathbf{b}];$$

together with

$$[\mathbf{a}\ \ \mathbf{b}\ \ \mathbf{a} \times \mathbf{b}] = (\mathbf{a} \times \mathbf{b}) \,.\, (\mathbf{a} \times \mathbf{b}) = |\,\mathbf{a} \times \mathbf{b}\,|^2.$$

Denoting magnitudes of vectors **a** and **b** by a and b respectively, above vectors simplify as

$$^{-1}\mathbf{a} = \{b^2\,\mathbf{a} - (\mathbf{b}\,.\,\mathbf{a})\,\mathbf{b}\} / |\,\mathbf{a} \times \mathbf{b}\,|^2,$$

$$^{-1}\mathbf{b} = \{\,a^2\,\mathbf{b} - (\mathbf{b}\,.\,\mathbf{a})\,\mathbf{a}\} / |\,\mathbf{a} \times \mathbf{b}\,|^2,$$

and

$$^{-1}(\mathbf{a} \times \mathbf{b}) = (\mathbf{a} \times \mathbf{b}) / |\,\mathbf{a} \times \mathbf{b}\,|^2. \ //$$

Example 7.2. Find the reciprocal system of vectors

$$\mathbf{a} = -\,\hat{\mathbf{i}} + \hat{\mathbf{j}} + \hat{\mathbf{k}}\,, \qquad \mathbf{b} = \hat{\mathbf{i}} - \hat{\mathbf{j}} + \hat{\mathbf{k}} \quad \text{and} \qquad \mathbf{c} = \hat{\mathbf{i}} + \hat{\mathbf{j}} - \hat{\mathbf{k}}\,,$$

where the vectors $\hat{\mathbf{i}}$, $\hat{\mathbf{j}}$, $\hat{\mathbf{k}}$ form the basis of the space.

Solution. Following Eqs. (7.1), and putting from Eqs. (4.18), we get

$$^{-1}\mathbf{a} = \{\,(\hat{\mathbf{i}} - \hat{\mathbf{j}} + \hat{\mathbf{k}}\,) \times (\hat{\mathbf{i}} + \hat{\mathbf{j}} - \hat{\mathbf{k}}\,)\,\} / [\mathbf{a}\ \ \mathbf{b}\ \ \mathbf{c}] = 2\,(\hat{\mathbf{j}} + \hat{\mathbf{k}}\,) / [\mathbf{a}\ \ \mathbf{b}\ \ \mathbf{c}],$$

$$^{-1}\mathbf{b} = \{\,(\hat{\mathbf{i}} + \hat{\mathbf{j}} - \hat{\mathbf{k}}\,) \times (-\,\hat{\mathbf{i}} + \hat{\mathbf{j}} + \hat{\mathbf{k}}\,)\,\} / [\mathbf{a}\ \ \mathbf{b}\ \ \mathbf{c}] = 2\,(\hat{\mathbf{k}} + \hat{\mathbf{i}}\,) / [\mathbf{a}\ \ \mathbf{b}\ \ \mathbf{c}],$$

and

$$^{-1}\mathbf{c} = \{(-\,\hat{\mathbf{i}} + \hat{\mathbf{j}} + \hat{\mathbf{k}}\,) \times (\hat{\mathbf{i}} - \hat{\mathbf{j}} + \hat{\mathbf{k}}\,)\} / [\mathbf{a}\ \ \mathbf{b}\ \ \mathbf{c}] = 2\,(\hat{\mathbf{i}} + \hat{\mathbf{j}}) / [\mathbf{a}\ \ \mathbf{b}\ \ \mathbf{c}],$$

together with

$$[\mathbf{a}\ \ \mathbf{b}\ \ \mathbf{c}] = \begin{vmatrix} -1 & 1 & 1 \\ 1 & -1 & 1 \\ 1 & 1 & -1 \end{vmatrix} = 4, \quad \text{by Eq. (5.3).}$$

Hence, we get

$$^{-1}\mathbf{a} = (\hat{\mathbf{j}} + \hat{\mathbf{k}}\,) / 2, \quad ^{-1}\mathbf{b} = (\hat{\mathbf{k}} + \hat{\mathbf{i}}\,) / 2 \quad \text{and} \quad ^{-1}\mathbf{c} = (\hat{\mathbf{i}} + \hat{\mathbf{j}}) / 2. \ //$$

§ 8. Vector equations

Here we present some equations involving vectors and seek their solutions mainly based upon the concept of Cor. 6.1. The method is illustrated by mean of few Examples.

Example 8.1. Solve the following equation for the vector **r** involving two given vectors **a** and **b**:

$$\mathbf{r} \times \mathbf{a} = \mathbf{b}. \tag{8.1}$$

Solution. The vector **a** × **b** being orthogonal to either of **a** and **b**, the vectors **a**, **b** and **a** × **b** are non-coplanar. Hence, in view of Cor. 6.1, any vector **r** in the space may be linearly expressed in terms of these non-coplanar vectors:

$$\mathbf{r} = u\,\mathbf{a} + v\,\mathbf{b} + w\,(\mathbf{a} \times \mathbf{b}), \tag{8.2}$$

where u, v, w are some scalar coefficients to be determined so as to the vector **r** satisfies Eq. (8.1). Forming cross product of Eq. (8.2) with vector **a** and putting from Eqs. (4.18) and (5.12). there results

$$\mathbf{r} \times \mathbf{a} = v\,(\mathbf{b} \times \mathbf{a}) + w\,(\mathbf{a} \times \mathbf{b}) \times \mathbf{a} = v\,(\mathbf{b} \times \mathbf{a}) + w\,\{(\mathbf{a}.\mathbf{a})\,\mathbf{b} - (\mathbf{b}.\mathbf{a})\,\mathbf{a}\},$$

or,

$$(1 - w\,a^2)\,\mathbf{b} + w\,(\mathbf{b}\,.\,\mathbf{a})\,\mathbf{a} + v\,(\mathbf{a} \times \mathbf{b}) = \mathbf{0}, \quad \text{by Eqs. (4.17) and (8.1).}$$

Vectors **a**, **b** and **a** × **b** being linearly independent, their scalar coefficients in above vector equation vanish identically in view of Defn. 1.3.2:

$$1 - w\,a^2 = w\,(\mathbf{b}\,.\,\mathbf{a}) = v = 0,$$

i.e.

$$v = 0, \qquad w = 1/a^2 \qquad \text{and} \qquad \mathbf{b}\,.\,\mathbf{a} = 0. \tag{8.3}$$

Putting for these values in Eq. (8.1), the desired solution is obtained:

$$\mathbf{r} = u\,\mathbf{a} + (1/a^2)\,(\mathbf{a} \times \mathbf{b}),$$

subject to the condition vide Eq. (8.3) requiring orthogonality of vectors **a**, **b**. //

Example 8.2. Solve the simultaneous equations for the vector **r** involving three given vectors **a**, **b** and **c** :

$$\mathbf{r} \times \mathbf{a} = \mathbf{a} \times \mathbf{b} \quad (8.4\text{a}); \quad \text{and} \quad \mathbf{r}\,.\,\mathbf{c} = 0. \tag{8.4b}$$

Solution. As in previous example, a vector **r** in the space may be linearly expressed in terms of the non-coplanar vectors **a**, **b** and **a** × **b**. Substituting for **r** from Eq. (8.2), Eq. (8.4a) yields

$$\mathbf{r} \times \mathbf{a} = v\,(\mathbf{b} \times \mathbf{a}) + w\,\{(\mathbf{a.a})\,\mathbf{b} - (\mathbf{b.a})\,\mathbf{a}\},$$

or, by Eq. (8.4a)

$$(1 + v)\,(\mathbf{a} \times \mathbf{b}) - w\,a^2\,\mathbf{b} + w\,(\mathbf{b}\,.\,\mathbf{a})\,\mathbf{a} = \mathbf{0},$$

by Eqs. (4.17), (4.18) and (8.1). For the same reasons as in previous example, above equation causes vanishing of the scalar coefficients:

$$1 + v = -w\,a^2 = w\,(\mathbf{b}\,.\,\mathbf{a}) = 0 \Rightarrow v = -1,\ w = 0. \qquad (8.5)$$

Further, putting for **r** from Eq. (8.2) in Eqs. (8.4b), there follows:

$$0 = u\,(\mathbf{a}\,.\,\mathbf{c}) - (\mathbf{b}\,.\,\mathbf{c}) \quad \Rightarrow \quad u = (\mathbf{b}\,.\,\mathbf{c})\,/\,(\mathbf{a}\,.\,\mathbf{c}), \qquad (8.6)$$

for Eqs. (8.4) and (8.5). Hence, Eq. (8.2) provides the solution

$$\mathbf{r} = u\,\mathbf{a} - \mathbf{b} = \{(\mathbf{b}\,.\,\mathbf{c})\,/\,(\mathbf{a}\,.\,\mathbf{c})\}\,\mathbf{a} - \mathbf{b}, \qquad (8.7)$$

by Eq. (8.6); provided **a. c** ≠ 0, i.e. they are not orthogonal to each other.

Aliter. Eq. (8.4a) can be re-written as

$$(\mathbf{r} + \mathbf{b}) \times \mathbf{a} = \mathbf{0} \quad \Rightarrow \quad \mathbf{r} + \mathbf{b} \text{ is parallel to } \mathbf{a},$$

i.e.

$$\mathbf{r} + \mathbf{b} = u\,\mathbf{a} \quad \Rightarrow \quad \mathbf{r} = u\,\mathbf{a} - \mathbf{b},$$

as in Eq. (8.7). //

Example 8.3. Find a condition for consistency of equations (8.1) and

$$\mathbf{r} \times \mathbf{c} = \mathbf{d}. \qquad (8.8)$$

Also find a common solution of the equations when the above condition is satisfied.

Solution. Assuming linear independence of the vectors **a**, **b**, **c** any vector **r** in the space may be a linear combination of these vectors and there exists a relation like Eq. (8.2):

$$\mathbf{r} = u\,\mathbf{a} + v\,\mathbf{b} + w\,\mathbf{c}. \qquad (8.9)$$

Substituting this value of **r** in Eqs. (8.1), and (8.8), we get

$$\mathbf{r} \times \mathbf{a} = v\,(\mathbf{b} \times \mathbf{a}) + w\,(\mathbf{c} \times \mathbf{a}) = \mathbf{b}, \qquad (8.10)$$

and

$$\mathbf{r} \times \mathbf{c} = u\,(\mathbf{a} \times \mathbf{c}) + v\,(\mathbf{b} \times \mathbf{c}) = \mathbf{d}. \qquad (8.11)$$

Forming dot products of these equations with vectors **c** and **a** respectively and putting from Eqs. (5.7) – (5.9), there follow two values of coefficient v:

$$v\ [\mathbf{b}\ \mathbf{a}\ \mathbf{c}] = \mathbf{b.c} \Rightarrow v = (\mathbf{b.c}) / [\mathbf{b}\ \mathbf{a}\ \mathbf{c}] = -(\mathbf{b.c}) / [\mathbf{a}\ \mathbf{b}\ \mathbf{c}],$$

and

$$v\ [\mathbf{b}\ \mathbf{c}\ \mathbf{a}] = \mathbf{d.a} \Rightarrow v = (\mathbf{d.a}) / [\mathbf{b}\ \mathbf{c}\ \mathbf{a}] = (\mathbf{a.d}) / [\mathbf{a}\ \mathbf{b}\ \mathbf{c}].$$

For consistency of given equations these values must be the same causing

$$\mathbf{a} \,.\, \mathbf{d} = -\,\mathbf{b} \,.\, \mathbf{c},$$

as the necessary condition. To have a common solution we now form the dot products of Eqs. (8.10) and (8.11) with vector **b**:

$$w\ [\mathbf{b}\ \mathbf{c}\ \mathbf{a}] = \mathbf{b} \,.\, \mathbf{b} \Rightarrow w = b^2 / [\mathbf{b}\ \mathbf{c}\ \mathbf{a}] = b^2 / [\mathbf{a}\ \mathbf{b}\ \mathbf{c}],$$

and

$$u\ [\mathbf{b}\ \mathbf{a}\ \mathbf{c}] = \mathbf{b} \,.\, \mathbf{d} \Rightarrow u = (\mathbf{b.d}) / [\mathbf{b}\ \mathbf{a}\ \mathbf{c}] = -(\mathbf{b.d}) / [\mathbf{a}\ \mathbf{b}\ \mathbf{c}].$$

Putting for the coefficients u, v, w in Eq. (8.9) the common solution of given equations is obtained:

$$\mathbf{r} = \{-(\mathbf{b} \,.\, \mathbf{d})\,\mathbf{a} + (\mathbf{a} \,.\, \mathbf{d})\,\mathbf{b} + b^2\,\mathbf{c}\} / [\mathbf{a}\ \mathbf{b}\ \mathbf{c}]. \ //$$

§ 9. Problem set

9.1. Show that the vectors $\mathbf{a} \times (\mathbf{b} \times \mathbf{c})$, $\mathbf{b} \times (\mathbf{c} \times \mathbf{a})$, $\mathbf{c} \times (\mathbf{a} \times \mathbf{b})$ are coplanar.

[**Hint:** Use Eq. (5.12) to split the scalar triple products and show that their sum vanishes identically.]

9.2. Show that the vector equation $k\,\mathbf{r} + \mathbf{r} \times \mathbf{a} = \mathbf{b}$ has a solution

$$(k^2 + a^2)\,\mathbf{r} = \{\,(\mathbf{a.\,b}) / k\,\}\,\mathbf{a} + k\,\mathbf{b} + \mathbf{a} \times \mathbf{b}. \qquad (9.1)$$

[**Hint:** Follow the method given in Example 9.1.]

9.3. Any vector **r** in the space satisfies the equations

$$\mathbf{r}\,.\,\mathbf{a} = l, \quad \mathbf{r}\,.\,\mathbf{b} = m, \quad \mathbf{r}\,.\,\mathbf{c} = p \quad \text{and} \quad \mathbf{r}\,.\,\mathbf{d} = q, \qquad (9.2)$$

then derive the relation

$$l\,[\mathbf{b}\;\;\mathbf{c}\;\;\mathbf{d}] + m\,[\mathbf{c}\;\;\mathbf{a}\;\;\mathbf{d}] + p\,[\mathbf{a}\;\;\mathbf{b}\;\;\mathbf{d}] = q\,[\mathbf{a}\;\;\mathbf{b}\;\;\mathbf{c}].$$

[**Hint:** Taking the vector **r** as a linear combination of non-coplanar vectors **b** × **c**, **c** × **a** and **a** × **b**:

$$\mathbf{r} = u\,(\mathbf{b}\times\mathbf{c}) + v\,(\mathbf{c}\times\mathbf{a}) + w\,(\mathbf{a}\times\mathbf{b}),$$

and putting it in given Eqs. (9.2), we evaluate the coefficients u, v, w.]

9.4. Establish the identity

$$[\mathbf{a}\;\;\mathbf{b}\;\;\mathbf{c}]^2 = \begin{vmatrix} \mathbf{a}\,.\,\mathbf{a} & \mathbf{a}\,.\,\mathbf{b} & \mathbf{a}\,.\,\mathbf{c} \\ \mathbf{b}\,.\,\mathbf{a} & \mathbf{b}\,.\,\mathbf{b} & \mathbf{b}\,.\,\mathbf{c} \\ \mathbf{c}\,.\,\mathbf{a} & \mathbf{c}\,.\,\mathbf{b} & \mathbf{c}\,.\,\mathbf{c} \end{vmatrix}, \qquad (9.3)$$

for any three vectors **a**, **b**, **c**. If these vectors form three adjacent sides of a cuboid the above formula determines square of the volume of the cuboid.

[**Hint:** Use Eq. (5.3) and multiply the two determinants. Next, use Eq. (5.5).]

9.5. Given three non-coplanar vectors **a**, **b**, **c** any vector **r** in the space can be expressed as

$$\mathbf{r} = (\Delta_1/\Delta)\,\mathbf{a} - (\Delta_2/\Delta)\,\mathbf{b} + (\Delta_3/\Delta)\,\mathbf{c},$$

where

$$\Delta_1 = \begin{vmatrix} \mathbf{r}\,.\,\mathbf{a} & \mathbf{r}\,.\,\mathbf{b} & \mathbf{r}\,.\,\mathbf{c} \\ \mathbf{b}\,.\,\mathbf{a} & \mathbf{b}\,.\,\mathbf{b} & \mathbf{b}\,.\,\mathbf{c} \\ \mathbf{c}\,.\,\mathbf{a} & \mathbf{c}\,.\,\mathbf{b} & \mathbf{c}\,.\,\mathbf{c} \end{vmatrix}, \Delta_2 = \begin{vmatrix} \mathbf{r}\,.\,\mathbf{a} & \mathbf{r}\,.\,\mathbf{b} & \mathbf{r}\,.\,\mathbf{c} \\ \mathbf{a}\,.\,\mathbf{a} & \mathbf{a}\,.\,\mathbf{b} & \mathbf{a}\,.\,\mathbf{c} \\ \mathbf{c}\,.\,\mathbf{a} & \mathbf{c}\,.\,\mathbf{b} & \mathbf{c}\,.\,\mathbf{c} \end{vmatrix}, \Delta_3 = \begin{vmatrix} \mathbf{r}\,.\,\mathbf{a} & \mathbf{r}\,.\,\mathbf{b} & \mathbf{r}\,.\,\mathbf{c} \\ \mathbf{a}\,.\,\mathbf{a} & \mathbf{a}\,.\,\mathbf{b} & \mathbf{a}\,.\,\mathbf{c} \\ \mathbf{b}\,.\,\mathbf{a} & \mathbf{b}\,.\,\mathbf{b} & \mathbf{b}\,.\,\mathbf{c} \end{vmatrix},$$

and Δ is the determinant in Eq. (9.3).

[**Hint:** Taking $\mathbf{r}$ as a linear sum of $\mathbf{a}$, $\mathbf{b}$, $\mathbf{c}$ as per Eq. (8.9), forming its dot products with the vectors $\mathbf{a}$, $\mathbf{b}$, $\mathbf{c}$ we eliminate the unknown coefficients u, v, w from Eq. (8.9) by means three equations so obtained.]

9.6. If the vectors $\hat{\mathbf{i}}$, $\hat{\mathbf{j}}$, $\hat{\mathbf{k}}$ form a self-reciprocal base show that they are the unit vectors of the original system.

[**Hint:** The vectors of the reciprocal system of $\hat{\mathbf{i}}$, $\hat{\mathbf{j}}$, $\hat{\mathbf{k}}$ are derived vide Eqs. (7.6) which, as per hypothesis, are the same. Further, the vectors of two systems satisfy equations analogous to Eqs. (7.2) and (7.3); which justify the orthonormal character of the given vectors.]

9.7. Show that the vector $\mathbf{r}$ satisfying equations

$$\mathbf{a}\,.\,\mathbf{r} = 1, \quad \mathbf{b}\,.\,\mathbf{r} = \mathbf{c}\,.\,\mathbf{r} = 0, \quad \text{but} \qquad \mathbf{b} \times \mathbf{c} \neq \mathbf{0},$$

is given by

$$\mathbf{r} = (\mathbf{b} \times \mathbf{c}) \,/\, [\mathbf{a}\ \ \mathbf{b}\ \ \mathbf{c}].$$

[**Hint:** Take $\mathbf{r} = u\,\mathbf{a} + v\,\mathbf{c} + w\,(\mathbf{b} \times \mathbf{c})$ and put it in the given equations to evaluate $u = v = 0$ and $w = 1 \,/\, [\mathbf{a}\ \ \mathbf{b}\ \ \mathbf{c}]$.]

9.8. Show that the relations

$$\mathbf{a}\,.\,\mathbf{a}' = \mathbf{b}\,.\,\mathbf{b}' = \mathbf{c}\,.\,\mathbf{c}' = 1,$$

$$\mathbf{a}\,.\,\mathbf{b}' = \mathbf{a}\,.\,\mathbf{c}' = \mathbf{b}\,.\,\mathbf{a}' = \mathbf{b}\,.\,\mathbf{c}' = \mathbf{c}\,.\,\mathbf{a}' = \mathbf{c}\,.\,\mathbf{b}' = 0,$$

conclude that the two systems of vectors $\mathbf{a}$, $\mathbf{b}$, $\mathbf{c}$ and $\mathbf{a}'$, $\mathbf{b}'$, $\mathbf{c}'$ form reciprocal systems of vectors.

[**Hint:** Compare the given equations with Eqs. (7.2) and (7.3) characterizing the vectors of reciprocal systems.]

9.9. Solve the system of equations

$$\mathbf{a}\,.\,\mathbf{r} = m \qquad (9.4); \qquad \mathbf{b} \times \mathbf{r} = \mathbf{c}. \qquad (9.5)$$

[**Hint:** The vector $\mathbf{b} \times \mathbf{r}$, i.e. $\mathbf{c}$ being orthogonal to either of $\mathbf{b}$ and $\mathbf{r}$, there hold

$$\mathbf{b}\,.\,\mathbf{c} = \mathbf{r}\,.\,\mathbf{c} = 0. \qquad (9.6)$$

Thus, if the vector **a** makes angle θ with **b**, it makes angle 90° – θ with **c**. Therefore,

$$\mathbf{a}\,.\,\mathbf{b} = ab\cos\theta, \qquad \mathbf{a}\,.\,\mathbf{c} = ac\cos(90^\circ - \theta) = ac\sin\theta. \qquad (9.7)$$

Taking $\mathbf{r} = u\,\mathbf{a} + v\,\mathbf{b} + w\,(\mathbf{a}\times\mathbf{b})$ and putting it in the given equations, there follow:

$$m = u\,a^2 + v\,(\mathbf{a}\,.\,\mathbf{b}), \qquad (9.8)$$

$$\mathbf{c} = u\,(\mathbf{b}\times\mathbf{a}) + w\,\mathbf{b}\times(\mathbf{a}\times\mathbf{b}) = u\,(\mathbf{b}\times\mathbf{a}) + w\,\{b^2\,\mathbf{a} - (\mathbf{b}\,.\,\mathbf{a})\,\mathbf{b}\}, \qquad (9.9)$$

by Eqs. (4.18) and (5.12). The dot product of Eq. (9.9) with vectors **a** and **c**, for Eqs. (5.8), (9.6) and (9.7) determines

$$ac\sin\theta = w\,\{a^2 b^2 - (\mathbf{a.b})^2\} = w\,a^2b^2\sin^2\theta$$

⇒

$$w = c\,/\,ab^2\sin\theta; \qquad (9.10)$$

and

$$c^2 = u\,[\mathbf{c}\;\;\mathbf{b}\;\mathbf{a}] + wb^2\,(\mathbf{a}\,.\,\mathbf{c}) = u\,[\mathbf{c}\;\mathbf{b}\;\mathbf{a}] + c^2 \;\Rightarrow\; u = 0, \qquad (9.11)$$

by Eqs. (9.7) and (9.10). Hence, Eq. (9.8) determines $v = m\,/\,\mathbf{a.b}$.

Also, for Eqs. (5.9) and (9.11), the cross product of Eq. (9.9) with **a** yields

$$\mathbf{a}\times\mathbf{c} = -w\,(\mathbf{b}\,.\,\mathbf{a})\,(\mathbf{a}\times\mathbf{b}) \Rightarrow w\,(\mathbf{a}\times\mathbf{b}) = -(\mathbf{a}\times\mathbf{c})\,/\,(\mathbf{a}\,.\,\mathbf{b}).$$

Thus, the solution is $\mathbf{r} = \{m\,\mathbf{b} - (\mathbf{a}\times\mathbf{c})\}\,/\,(\mathbf{a}\,.\,\mathbf{b})$, if $\mathbf{a}\,.\,\mathbf{b} \neq 0$.]

9.10. Derive the formula

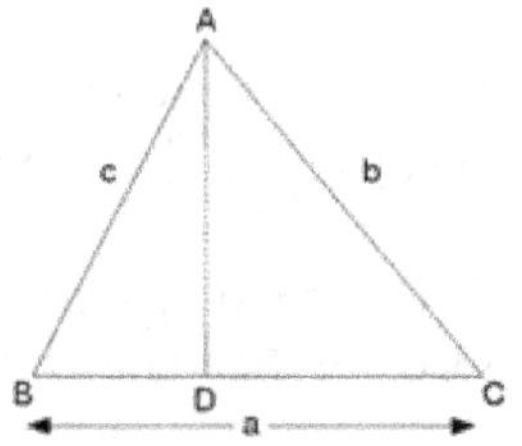

$$\Delta = \sqrt{\{s\,(s-a)\,(s-b)\,(s-c)\}}, \qquad (9.12)$$

for the area of a plane triangle ABC represented by the vectors **a**, **b**, **c** acting along its sides of lengths a, b, c , where $2s = a + b + c$ is the perimeter of the triangle.

[**Hint:** Area of the triangle ABC is (BC) (AD)/2 = $(ac/2)\sin B$, which is half the magnitude of $\mathbf{a}\times\mathbf{c}$ as per definition of the cross product.]

CHAPTER 2

VECTOR CALCULUS (DERIVATION)

§ 1. Vector functions of a single parameter and their derivation

Let $\mathbf{a}(t)$ and $\mathbf{a}(t + \delta t)$ be two neighbouring values of a vector function of a scalar parameter $t \varepsilon$ R. In analogy with the Chapter 4 of [11], we define the derivative of $\mathbf{a}(t)$ w.r.t. t by

$$d\,\mathbf{a} / dt \equiv \lim_{\delta t \to 0} \{\mathbf{a}(t + \delta t) - \mathbf{a}(t)\} / \delta t, \tag{1.1}$$

which is also a vector. Sometimes, especially when $\mathbf{a}(t)$ traces out a curve $\mathbf{r} = \mathbf{r}(t)$ and t is the time parameter its derivative w.r.t. t is also denoted by $\dot{\mathbf{r}}$ giving the velocity vector to the curve at the point P $(\mathbf{r})$. In analogy with the derivation rules of scalar functions there also hold the following rules for the derivation of vector functions.

Theorem 1.1. Let $\mathbf{a}$ and $\mathbf{b}$ be two vector functions and u a scalar function of the same scalar parameter t. There hold the following rules for derivation of their various combinations:

$$d(\mathbf{a} \pm \mathbf{b}) / dt = d\,\mathbf{a} / dt \pm d\,\mathbf{b} / dt, \tag{1.2}$$

$$d(u\,\mathbf{a}) / dt = u\,(d\,\mathbf{a} / dt) + (du / dt)\,\mathbf{a}, \tag{1.3}$$

$$d(\mathbf{a} \,.\, \mathbf{b}) / dt = (d\,\mathbf{a} / dt) .\, \mathbf{b} + \mathbf{a} \,.\, (d\,\mathbf{b} / dt), \tag{1.4}$$

$$d(\mathbf{a} \times \mathbf{b}) / dt = (d\,\mathbf{a} / dt) \times \mathbf{b} + \mathbf{a} \times (d\,\mathbf{b} / dt), \tag{1.5}$$

Proof of these results being simple is left to the reader. It may be noted that the order of vectors in the derivation of cross product of vectors must be preserved because of skew-symmetric property of the cross product.

Corollary 1.1. The derivative of any constant vector $\mathbf{c}$ w.r.t. any (scalar) parameter becomes a null vector.

$$d\,\mathbf{c} / dt = \mathbf{0}\,. \tag{1.6}$$

Theorem 1.2. For a vector **a** (t) with components a_1, a_2, a_3 along the rectangular Cartesian coordinate axes, its derivative w.r.t. t is given by

$$d\mathbf{a} / dt = \hat{\mathbf{i}} \; (da_1 / dt) + \hat{\mathbf{j}} \; (da_2 / dt) + \hat{\mathbf{k}} \; (da_3 / dt), \tag{1.7}$$

where $\hat{\mathbf{i}}$, $\hat{\mathbf{j}}$, $\hat{\mathbf{k}}$ are the unit vectors along the coordinate axes.

Proof. Forming the derivative of the vector

$$\mathbf{a}\,(t) = a_1\,(t)\,\hat{\mathbf{i}} + a_2\,(t)\,\hat{\mathbf{j}} + a_3\,(t)\,\hat{\mathbf{k}}, \tag{1.8}$$

and applying the Eqs. (1.3) and (1.6) we immediately get the result. //

Theorem 1.3. A vector **a** (t) is of constant magnitude if there holds the necessary and sufficient condition

$$\mathbf{a} \,.\, (d\mathbf{a} / dt) = \mathbf{0}. \tag{1.9}$$

Proof. A vector of constant magnitude a satisfies

$$\mathbf{a} \,.\, \mathbf{a} = \mathbf{a}^2 = a^2 \text{ (constant)}. \tag{1.10}$$

Its derivation for Eqs. (1.4) and (1.6) yields the condition (1.9).

On the other hand, for Eq. (1.10) and symmetric property of dot product of vectors, the Eq. (1.4) may be re-written as

$$d\,(a^2) / dt = d\,(\mathbf{a} \,.\mathbf{a}) / dt = (d\mathbf{a} / dt) \mathbf{.}\, \mathbf{a} + \mathbf{a} \,.\, (d\mathbf{a} / dt) = 2\mathbf{a} \,.\, (d\mathbf{a} / dt),$$

which vanishes when Eq. (1.9) holds. This establishes the sufficiency of the condition. //

Corollary 1.2. For a non-null vector **a** of constant magnitude its derivative $d\mathbf{a}/dt$ is either a null vector or orthogonal to **a**.

Proof is a necessary consequence of the Eq. (1.9). //

Theorem 1.4. A vector **a** (t) may have fixed direction if there holds the necessary and sufficient condition

$$\mathbf{a} \times (d\,\mathbf{a} / dt) = \mathbf{0}. \tag{1.11}$$

Proof. Two vectors **a** (t) and **a** ($t + \delta t$) at the neighbouring points P (t) and Q ($t + \delta t$) will be parallel if their cross product is zero:

$$\mathbf{a}\,(t) \times \mathbf{a}\,(t+\delta t) = \mathbf{0} \;\Rightarrow\; \mathbf{a}\,(t) \times [\{\mathbf{a}\,(t+\delta t) - \mathbf{a}\,(t)\}/\,\delta t\,] = \mathbf{0}.$$

Taking its limit as $\delta t \to 0$, and applying Eq. (1.1) we derive the result. //

Theorem 1.5. The (scalar) components velocity and acceleration of a particle P (r, θ) describing a plane curve $\mathbf{r} = \mathbf{r}\,(t)$ at any instant (of time) t are given by

	Velocity	Acceleration
Radial	$dr\,/\,dt$	$\ddot{r} - r\dot{\theta}^2$
Transverse	$r\dot{\theta}$	$2\dot{r}\,\dot{\theta} + r\ddot{\theta} = (1/r)\,d\,(r^2\dot{\theta}\,)\,/dt$
Tangential	$v = \lvert\,\dot{\mathbf{r}}\,\rvert = \dot{s}$ $= \sqrt{\{\,\dot{r}^2 + (r\dot{\theta}\,)^2\,\}}$	$dv/dt = \ddot{s}$ $= (1/2v)\,(d/dt)\,\{\,\dot{r}^2 + (r\dot{\theta}\,)^2\}$
Normal	0	$v^2\,/\,\rho$

where ρ is the radius if curvature of the path at P and θ is angle made by the radius vector $\overline{\mathrm{OP}} = \mathbf{r}$ with the initial line OA.

Proof. (i) Let $\overline{\mathrm{OP}} = \mathbf{r}$ and $\overline{\mathrm{OQ}} = \mathbf{r} + \delta\mathbf{r}$ be two neighbouring radii vectors on the path making angles θ and $\theta + \delta\theta$ to the initial line chosen along x-axis. Their joining chord $\overline{PQ}$ $= \delta\mathbf{r}$ becomes *tangential* vector $\mathbf{v} = \dot{\mathbf{r}} = d\mathbf{r}/dt$ to the path when the point Q tends to P. Let $\hat{\mathbf{i}}$, $\hat{\mathbf{j}}$ be the unit vectors along two rectangular coordinate axes Ox and Oy so that the radius vector **r** is expressible as

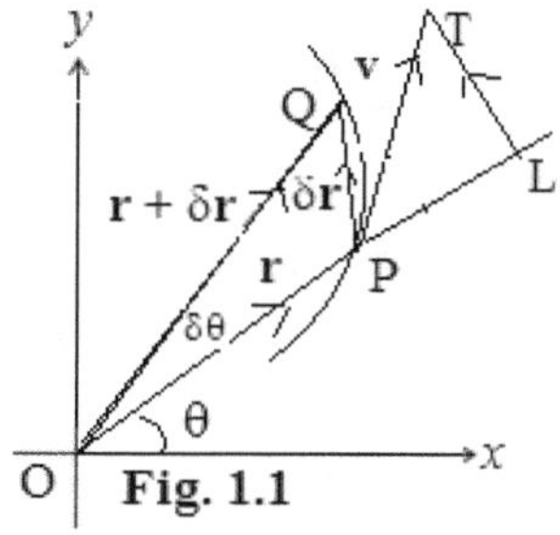

Fig. 1.1

$$\mathbf{r} = (\mathrm{OP.}\cos\theta)\;\hat{\mathbf{i}} + (\mathrm{OP.}\sin\theta)\;\hat{\mathbf{j}} = r\,(\,\hat{\mathbf{i}}\cos\theta + \hat{\mathbf{j}}\sin\theta), \qquad (1.12a)$$

with the unit vector along it

$$\hat{\mathbf{r}} = \mathbf{r} / r = (\hat{\mathbf{i}} \cos\theta + \hat{\mathbf{j}} \sin\theta). \qquad (1.12b)$$

Differentiating Eq. (1.12a) w.r.t. t we get the velocity vector

$$\mathbf{v} = \dot{\mathbf{r}} = (dr/dt)\hat{\mathbf{r}} + r(-\hat{\mathbf{i}} \sin\theta + \hat{\mathbf{j}} \cos\theta)(d\theta/dt) = \dot{r}\,\hat{\mathbf{r}} + r\dot{\theta}\,\hat{\mathbf{p}}, \qquad (1.13)$$

where the unit vector

$$\hat{\mathbf{p}} = -\hat{\mathbf{i}} \sin\theta + \hat{\mathbf{j}} \cos\theta \qquad (1.14)$$

is orthogonal to $\hat{\mathbf{r}}$:

$$\hat{\mathbf{r}} \cdot \hat{\mathbf{p}} = (\hat{\mathbf{i}} \cos\theta + \hat{\mathbf{j}} \sin\theta) \cdot (-\hat{\mathbf{i}} \sin\theta + \hat{\mathbf{j}} \cos\theta) = 0,$$

for orthonormal vectors $\hat{\mathbf{i}}$ and $\hat{\mathbf{j}}$ satisfying Eqs. (1.4.4).

Definition 1.1. The direction orthogonal to radius vector is called a *transverse* direction.

Thus, the (scalar) coefficients of $\hat{\mathbf{r}}$ and $\hat{\mathbf{p}}$ in Eq. (1.13) form the radial and transverse components of the velocity vector $\mathbf{v}$, which is directed towards the tangent PT. Its magnitude forms the tangential velocity while its normal component (being orthogonal to it) is zero.

(ii) Also, in limiting case the length of chord PQ equals arc-length PQ = δs implying $|\,\dot{\mathbf{r}}\,| = \dot{s} = v$.

(iii) Further derivation of Eq. (1.13) provides the acceleration vector

$$\mathbf{a} = d\mathbf{v} / dt = \ddot{\mathbf{r}} \equiv d^2\mathbf{r} / dt^2 = (\ddot{r} - r\dot{\theta}^2)\,\hat{\mathbf{r}} + (2\dot{r}\,\dot{\theta} + r\ddot{\theta})\,\hat{\mathbf{p}}. \qquad (1.15)$$

The coefficients of $\hat{\mathbf{r}}$ and $\hat{\mathbf{p}}$ in above equation form the radial and transverse components of the acceleration vector. The latter can also be written as $(1/r)\, d\,(r^2\dot{\theta}\,) / dt$.

(iv) For

$$\mathbf{v} = d\mathbf{r} / dt = (d\mathbf{r} / ds)\,(ds / dt) = \hat{\mathbf{t}}\,\dot{s}, \qquad (1.16)$$

where $\hat{\mathbf{t}}$ is the unit tangent vector to the curve. Accordingly, the acceleration vector is

$$\mathbf{a} = d\mathbf{v} / dt = (d\,\hat{\mathbf{t}} / ds)\,\dot{s}^2 + \hat{\mathbf{t}}\,\ddot{s}. \qquad (1.17)$$

In view of Corollary 1.2, the vector $(d\hat{\mathbf{t}}\ /\ ds)$ is orthogonal to the tangent vector. Hence, it is along the normal to the curve. Being the rate of change of the direction of the tangent, it defines the curvature vector, say $\kappa\overline{\mathbf{n}} = \overline{\mathbf{n}} / \rho$. Thus, the normal acceleration is $\dot{s}^2 / \rho = v^2 / \rho$ while the coefficient $\ddot{s}$ of the tangent vector provides the tangential acceleration. //

Corollary 1.3. If a particle describes a circle of radius, say a (constant) the results of above theorem reduce to

	Velocity	Acceleration
Radial	0	$-a\dot{\theta}^2$
Transverse	$a\dot{\theta}$	$a\ddot{\theta}$
Tangential	$v = \lvert \dot{\mathbf{r}} \rvert = \dot{s} = a\dot{\theta}$	$dv/dt = \ddot{s} = a\ddot{\theta}$
Normal	0	v^2 / a

Proof. Radius of a circle being constant its derivatives are zero and the results of previous theorem reduce to above form. //

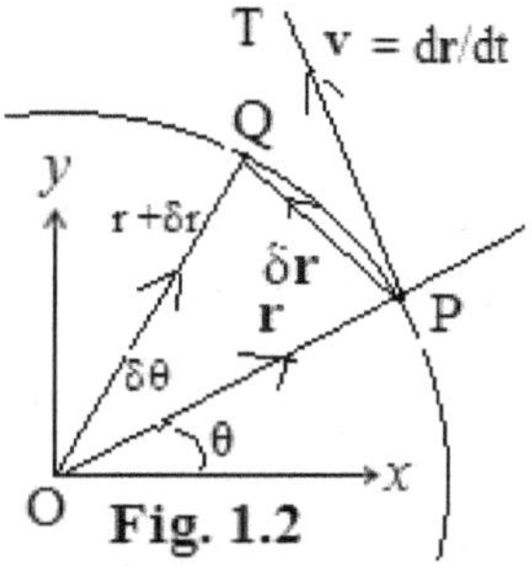

Fig. 1.2

Corollary 1.4. Further, if the angular velocity $\dot{\theta}$ also remains constant, say ω, the results in above Corollary further simplify as

	Velocity	Acceleration
Radial	0	$-\omega^2 a$
Transverse	ωa	0
Tangential	$v = \lvert \dot{\mathbf{r}} \rvert = \dot{s} = \omega a$	0
Normal	0	$\omega^2 a$

Note 1.1. The negative sign in the radial acceleration shows its direction directed towards the centre.

Example 1.1. For a vector $\mathbf{r} = \mathbf{a} \cos \omega t + \mathbf{b} \sin \omega t$, where $\mathbf{a}$, $\mathbf{b}$, ω are constant, there hold the relations

$$\mathbf{r} \times (d\,\mathbf{r}/dt) = \omega\, \mathbf{a} \times \mathbf{b}, \qquad \text{and} \qquad d^2\,\mathbf{r} / dt^2 + \omega^2\,\mathbf{r} = \mathbf{0}.$$

Solution. Differentiating the vector $\mathbf{r}$ and taking into account the constant nature of $\mathbf{a}$, $\mathbf{b}$ and ω, we get

$$d\mathbf{r} / dt = \omega\,(-\,\mathbf{a} \sin \omega t + \mathbf{b} \cos \omega t). \tag{1.18}$$

Its cross product with $\mathbf{r}$, for vanishing characteristics of cross product of a vector with itself, yields

$$\mathbf{r} \times (d\mathbf{r} / dt) = \omega\, \{(\mathbf{a} \times \mathbf{b}) \cos^2 \omega t - (\mathbf{b} \times \mathbf{a}) \sin^2 \omega t\} = \omega\, \mathbf{a} \times \mathbf{b},$$

where skew-symmetric property of cross product is also used.

Further derivation of Eq. (1.18) w.r.t. t easily establishes the result. //

Example 1.2. For a unit vector $\hat{\mathbf{u}}$ show that

$$|\,\hat{\mathbf{u}} \times (d\,\hat{\mathbf{u}} / dt)| = |\,d\,\hat{\mathbf{u}} / dt\,|.$$

Solution. In view of Corollary 1.2, the derivative of the vector $\hat{\mathbf{u}}$ is orthogonal to $\hat{\mathbf{u}}$. Hence, the magnitude of the cross product vector depends upon the magnitude of the derivative vector. //

§ 2. Partial derivation of vector functions

Let $\mathbf{a} = (x, y, z)$ be a vector function of three independent variables x, y, z. The limit, if it exists,

$$\lim_{\partial x \to 0}\ \{\mathbf{a}\,(x + \partial x\,, y, z) - \mathbf{a}\,(x, y, z)\} \,/\, \partial x$$

is called the *partial derivative* of $\mathbf{a}$ with respect to x and is denoted by $\partial\mathbf{a} / \partial x$. Similarly, the partial derivatives of $\mathbf{a}$ with respect to y and z can be defined. If the vector $\mathbf{a}$ has components (a_1, a_2, a_3), its partial derivatives can be obtained by

$$\partial\mathbf{a} / \partial x = \hat{\mathbf{i}}\ (\partial a_1 / \partial x) + \hat{\mathbf{j}}\ (\partial a_2 / \partial x) + \hat{\mathbf{k}}\ (\partial a_3 / \partial x), \text{ etc.} \tag{2.1}$$

We introduce a vector differential operator (called '*del*' or '*nabla*')

$$\nabla \equiv \hat{\mathbf{i}}\ (\partial / \partial x) + \hat{\mathbf{j}}\ (\partial / \partial y) + \hat{\mathbf{k}}\ (\partial / \partial z). \tag{2.2}$$

Definition 2.1. Given a scalar point function $f\ (x, y, z)$, the vector

$$\nabla f \equiv \hat{\mathbf{i}}\ (\partial f / \partial x) + \hat{\mathbf{j}}\ (\partial f / \partial y) + \hat{\mathbf{k}}\ (\partial f / \partial z). \tag{2.3}$$

is called the *gradient* of the function f.

For brevity, it is denoted by grad f and it has the components $\partial f / \partial x$, $\partial f / \partial y$, $\partial f / \partial z$ along the rectangular Cartesian coordinate axes Ox, Oy and Oz respectively.

2.1. Normal to a surface: We know that a functional relation

$$f\ (x, y, z) = c, \tag{2.4}$$

where c is an arbitrary constant (i.e. a fixed real number chosen arbitrarily) represents a family of surfaces in 3-dimensional Euclidean space. As three variables x, y, z are connected together no longer all of them can be independent to each other. At most (any) two of them can be chosen as independent while the remaining one depends on the independent ones. The number of independent variables determines the dimension of the locus represented by Eq. (2.4). If the function f is of degree one in the variables the surface becomes a plane otherwise a curved surface.

Let P (x, y, z) and Q $(x + dx, y + dy, z + dz)$ be two neighbouring points on the surface so that their coordinates satisfy Eq. (2.4):

$$f\ (x, y, z) = c, \qquad f\ (x + dx, y + dy, z + dz) = c,$$

$\Rightarrow$

$$df \equiv f\ (x + dx, y + dy, z + dz) - f(x, y, z) = 0,$$

or, by Taylor's theorem applied to the first order of approximation

$$(\partial f / \partial x)\ dx + (\partial f / \partial y)\ dy + (\partial f / \partial z)\ dz = 0, \tag{2.5a}$$

or, equivalently

$$(\partial f / \partial x, \partial f / \partial y, \partial f / \partial z) \,.\, (dx, dy, dz) \equiv \nabla f \,.\, d\mathbf{r} = 0. \tag{2.5b}$$

Thus, the two vectors ∇f and $d\mathbf{r}$ are orthogonal to each other. The vector $d\mathbf{r}$ in the limiting case, when Q → P, becomes *tangential* to the surface. Hence, the gradient vector ∇f, being orthogonal to $d\mathbf{r}$, becomes *normal* to the surface at P.

In the following, we define the directional derivative of a function along a unit vector $\hat{\mathbf{u}}$:

Definition 2.2. Let P (x, y, z) be a point on a plane surface represented by Eq. (2.4) and s measures the distance from P along a unit vector $\hat{\mathbf{u}}$ (taken on the plane). The derivative df / ds defines the directional derivative of f along $\hat{\mathbf{u}}$.

Theorem 2.1. The directional derivative of a function f at a point P (x, y, z) along a unit vector $\hat{\mathbf{u}}$ is also given by

$$df / ds = \hat{\mathbf{u}} . \nabla f. \tag{2.6}$$

Proof. As assumed earlier, s measures the distance from P along $\hat{\mathbf{u}}$, the vector $\hat{\mathbf{u}}$ is given by

$$\hat{\mathbf{u}} = \hat{\mathbf{i}}\ (dx / ds) + \hat{\mathbf{j}}\ (dy / ds) + \hat{\mathbf{k}}\ (dz / ds).$$

⇒

$$\hat{\mathbf{u}} . \nabla f = (dx / ds, dy / ds, dz / ds) \mathbf{.} (\partial f / \partial x, \partial f / \partial y, \partial f / \partial z)$$

$$= (\partial f / \partial x)\ (dx / ds) + (\partial f / \partial y)\ (dy / ds) + (\partial f / \partial z)\ (dz / ds) = df / ds. \ //$$

Corollary 2.1. Let $\hat{\mathbf{n}}$ be a unit normal vector to a plane surface represented by Eq. (1.4) in the increasing direction of f and n measures the distance from P along $\hat{\mathbf{n}}$. There holds the relation

$$\text{grad} f = (df / dn)\ \hat{\mathbf{n}}. \tag{2.7}$$

Proof. The vector grad f itself is normal to the surface, so we may have

$$\text{grad} f = \lambda\ \hat{\mathbf{n}}, \tag{2.8}$$

for some scalar λ. Its dot product with $\hat{\mathbf{n}}$, for Eq. (2.6), yields

$$df / dn \equiv \hat{\mathbf{n}} \mathbf{.}\ \text{grad} f = \lambda\ \hat{\mathbf{n}} \mathbf{.}\ \hat{\mathbf{n}} = \lambda,$$

which reduces Eq. (2.8) to the form of Eq. (2.7). //

Theorem 2.2. Given two scalar point functions $f(x, y, z)$ and $g(x, y, z)$, there hold the identities

$$\left.\begin{aligned} \text{grad}\,(f \pm g) &= \text{grad}\, f \pm \text{grad}\, g, \\ \text{grad}\,(fg) &= (\text{grad}\, f)\, g + f(\text{grad}\, g). \end{aligned}\right\} \quad (2.9)$$

Proof. (i) We have

$$\text{grad}\,(f \pm g) = \{\hat{\mathbf{i}}\,(\partial / \partial x) + \hat{\mathbf{j}}\,(\partial / \partial y) + \hat{\mathbf{k}}\,(\partial / \partial z)\}\,(f \pm g)$$

$$= \{\hat{\mathbf{i}}\,(\partial/\partial x) + \hat{\mathbf{j}}\,(\partial/\partial y) + \hat{\mathbf{k}}\,(\partial/\partial z)\}\, f \pm \{\hat{\mathbf{i}}\,(\partial/\partial x) + \hat{\mathbf{j}}\,(\partial/\partial y) + \hat{\mathbf{k}}\,(\partial/\partial z)\} g$$

$$= \text{grad}\, f \pm \text{grad}\, g.$$

(ii) $$\text{grad}\,(fg) = \{\hat{\mathbf{i}}\,(\partial / \partial x) + \hat{\mathbf{j}}\,(\partial / \partial y) + \hat{\mathbf{k}}\,(\partial / \partial z)\}(fg)$$

$$= \{\hat{\mathbf{i}}\,(\partial f / \partial x) + \hat{\mathbf{j}}\,(\partial f / \partial y) + \hat{\mathbf{k}}\,(\partial f / \partial z)\} g$$

$$+ f\{\hat{\mathbf{i}}\,(\partial g / \partial x) + \hat{\mathbf{j}}\,(\partial g / \partial y) + \hat{\mathbf{k}}\,(\partial g / \partial z)\} = (\text{grad}\, f)\, g + f(\text{grad}\, g). \; //$$

Note 2.1. The first identity in Eqs. (2.9) establishes the distributive property of the gradient operator over addition / subtraction of two scalar functions.

Example 2.1. Find the directional derivatives of a scalar point function $f(x, y, z)$ along the coordinate axes.

Solution. The unit vectors along the (rectangular Cartesian) coordinate axes Ox, Oy, Oz are $\hat{\mathbf{i}}$, $\hat{\mathbf{j}}$, $\hat{\mathbf{k}}$ respectively and x, y, z measure the distances of a point P (x, y, z) from the origin along these axes. So, by Eq. (2.6), the directional derivative of f along Ox axis is

$$\hat{\mathbf{i}} . \nabla f = \hat{\mathbf{i}} . \{\hat{\mathbf{i}}\,(\partial f / \partial x) + \hat{\mathbf{j}}\,(\partial f / \partial y) + \hat{\mathbf{k}}\,(\partial f / \partial z)\} = \partial f / \partial x.$$

Similarly, the directional derivatives of f along Oy and Oz are $\partial f / \partial y$ and $\partial f / \partial z$ respectively. //

Example 2.2. Let the position vector $\mathbf{r}$ of a point P (x, y, z) with respect to some origin O have magnitude r and m be any fixed real number. The gradient of scalar function r^m is given by

$$\text{grad}\, r^m = (m\, r^{m-2})\, \mathbf{r}. \tag{2.10}$$

Solution. By definition,

$$\text{grad}\, r^m = \{\hat{\mathbf{i}}\,(\partial / \partial x) + \hat{\mathbf{j}}\,(\partial / \partial y) + \hat{\mathbf{k}}\,(\partial / \partial z)\} r^m$$

$$= m\, r^{m-1}\, \{\hat{\mathbf{i}}\,(\partial / \partial x) + \hat{\mathbf{j}}\,(\partial / \partial y) + \hat{\mathbf{k}}\,(\partial / \partial z)\}\, r. \tag{2.11}$$

The magnitude r is given by Eq. (1.4.6). On differentiation with respect to x, it yields

$$r\,(\partial r / \partial x) = x \Rightarrow \partial r / \partial x = x / r.$$

Similarly,

$$\left.\begin{aligned} &\partial r / \partial y = y / r \quad \text{and} \quad \partial r / \partial z = z / r. \end{aligned}\right\} \tag{2.12}$$

Hence, Eq. (2.11) reduces to

$$\text{grad}\, r^m = m\, r^{m-2}\,(x\,\hat{\mathbf{i}} + y\,\hat{\mathbf{j}} + z\,\hat{\mathbf{k}}) = (m\, r^{m-2})\, \mathbf{r}, \quad \text{by Eq. (1.3.1).} \ //$$

Example 2.3. Show that

(i) $\{\text{grad} f(r)\} \times \mathbf{r} = \mathbf{0}$ (2.13); (ii) $\text{grad}\, |\,\mathbf{r}\,|^3 = (3\, r)\, \mathbf{r}$, (2.14)

and

(iii) $$\text{grad}\,(f / g) = \{(\text{grad} f)\, g - f\, (\text{grad}\, g)\}/g^2. \tag{2.15}$$

Solution. (i) $\text{grad} f(r) = \{\hat{\mathbf{i}}\,(\partial / \partial x) + \hat{\mathbf{j}}\,(\partial / \partial y) + \hat{\mathbf{k}}\,(\partial / \partial z)\} f(r)$

$$= \{\hat{\mathbf{i}}\,(\partial f / \partial x) + \hat{\mathbf{j}}\,(\partial f / \partial y) + \hat{\mathbf{k}}\,(\partial f / \partial z)\}$$

$$= (\partial f / \partial r)\, \{\hat{\mathbf{i}}\,(\partial r / \partial x) + \hat{\mathbf{j}}\,(\partial r / \partial y) + \hat{\mathbf{k}}\,(\partial r\, \partial z)\}$$

$$= (1/ r)\,(\partial f / \partial r)\,(x\,\hat{\mathbf{i}} + y\,\hat{\mathbf{j}} + z\,\hat{\mathbf{k}}) = (1/ r)\,(\partial f / \partial r)\, \mathbf{r},$$

by Eq. (2.12). Hence, its vector product with $\mathbf{r}$ becomes a null vector.

(ii) The result follows immediately from Eq. (2.10).

(iii) Treating f/g as a product of two scalar functions f and $1/g$, and applying the second identity in Eqs. (2.9), we get

$$\text{grad}\,(f/g) = (\text{grad}\,f)/g + f\,(\text{grad}\,g^{-1}). \tag{2.16}$$

But,

$$\text{grad}\,g^{-1} = \{\hat{\mathbf{i}}\,(\partial/\partial x) + \hat{\mathbf{j}}\,(\partial/\partial y) + \hat{\mathbf{k}}\,(\partial/\partial z)\}\,(g^{-1})$$

$$= (\partial g^{-1}/\partial g)\,\{\hat{\mathbf{i}}\,(\partial g/\partial x) + \hat{\mathbf{j}}\,(\partial g/\partial y) + \hat{\mathbf{k}}\,(\partial g/\partial z)\} = (-1/g^2)\,\text{grad}\,g.$$

Hence, Eq. (2.16) yields the result. //

Example 2.4. For any three constant vectors **a**, **b**, **c** and a fixed real number n, show that

(i) {grad $(\mathbf{r}\,.\,\mathbf{a}) = \mathbf{a}$ (2.17); **(ii)** grad $[\mathbf{r}\ \mathbf{a}\ \mathbf{b}] = \mathbf{a}\times\mathbf{b}$, (2.18)

and

(iii) $$\text{grad}\,|\,\mathbf{c}\times\mathbf{r}\,|^{n} = n\,|\,\mathbf{c}\times\mathbf{r}\,|^{n-2}\,(\mathbf{c}\times\mathbf{r}), \tag{2.19}$$

r, being the position vector of a point P (x, y, z).

Solution. (i) The dot product of vectors $\mathbf{r} = (x, y, z)$ and $\mathbf{a} = (a_1, a_2, a_3)$ being $\mathbf{r}.\,\mathbf{a} = a_1x + a_2\,y + a_3\,z$, we have

$$\text{grad}\,(\mathbf{r}.\,\mathbf{a}) = \{\hat{\mathbf{i}}\,(\partial/\partial x) + \hat{\mathbf{j}}\,(\partial/\partial y) + \hat{\mathbf{k}}\,(\partial/\partial z)\}(a_1\,x + a_2\,y + a_3\,z)$$

$$= a_1\,\hat{\mathbf{i}} + a_2\,\hat{\mathbf{j}} + a_3\,\hat{\mathbf{k}} = \mathbf{a}.$$

(ii) Scalar triple product of vectors is given by Eq. (1.5.3):

$$[\mathbf{r}\ \mathbf{a}\ \mathbf{b}] = \begin{vmatrix} x & y & z \\ a_1 & a_2 & a_3 \\ b_1 & b_2 & b_3 \end{vmatrix}$$

$$= (a_2\,b_3 - a_3\,b_2)\,x + (a_3\,b_1 - a_1\,b_3)\,y + (a_1\,b_2 - a_2\,b_1)\,z$$

$\Rightarrow$

$$\left.\begin{aligned} &\text{grad}\,[\mathbf{r}\ \mathbf{a}\ \mathbf{b}] = \{\hat{\mathbf{i}}\,(\partial/\partial x) + \hat{\mathbf{j}}\,(\partial/\partial y) + \hat{\mathbf{k}}\,(\partial/\partial z)\}\,[\mathbf{r}\ \mathbf{a}\ \mathbf{b}] \\ &= (a_2b_3 - a_3b_2)\,\hat{\mathbf{i}} + (a_3b_1 - a_1b_3)\,\hat{\mathbf{j}} + (a_1b_2 - a_2b_1)\,\hat{\mathbf{k}} = \mathbf{a}\times\mathbf{b}, \end{aligned}\right\} \tag{2.20}$$

by Eq. (1.4.19).

(iii) The vector **c** being constant, $\mathbf{c} \times \mathbf{r}$ depends on **r** only. Thus, applying Eq. (2.10), there follows the Eq. (2.19). //

Example 2.5. An electrostatic force is derived from the function $f(x, y) = x / (x^2 + y^2)$ in the direction of the grad f. Compute the force and evaluate its value at the point P (1, 1).

Solution. The function has partial derivatives

$$f_x = \partial f / \partial x = (y^2 - x^2) / (x^2 + y^2)^2,$$

$$f_y = \partial f / \partial y = -2xy / (x^2 + y^2)^2, \text{ and } f_z = 0.$$

Hence,

$$\nabla f = (f_x, f_y, f_z) = (y^2 - x^2, -2xy, 0) / (x^2 + y^2)^2,$$

and at P (1, 1) it is

$$(0, -2, 0) / 4 = (-1/2)\,\hat{\mathbf{j}}. \;//$$

§ 3. Divergence of a Vector

The position vector **r** of a point P (x, y, z) in terms of its components along the rectangular Cartesian coordinate axes is given by Eq. (1.3.1). Hence, its partial derivatives w.r.t. the coordinates are

$$\partial \mathbf{r} / \partial x = \hat{\mathbf{i}}, \quad \partial \mathbf{r} / \partial y = \hat{\mathbf{j}}, \quad \partial \mathbf{r} / \partial z = \hat{\mathbf{k}}. \tag{3.1}$$

It is seen in the previous Section that the vector differential operator ∇, given by Eq. (2.2), when applied over a scalar function defines the gradient vector. In contrast, if it is applied over a vector function, say **u**, under *dot* product it yields a scalar function called the *divergence* of (the vector) **u**, briefly denoted as div **u:**

$$\nabla \cdot \mathbf{u} = \{\hat{\mathbf{i}}\,(\partial / \partial x) + \hat{\mathbf{j}}\,(\partial / \partial y) + \hat{\mathbf{k}}\,(\partial / \partial z)\} \cdot \mathbf{u}$$

$$= \hat{\mathbf{i}} \cdot (\partial \mathbf{u} / \partial x) + \hat{\mathbf{j}} \cdot (\partial \mathbf{u} / \partial y) + \hat{\mathbf{k}} \cdot (\partial \mathbf{u} / \partial z). \tag{3.2}$$

Theorem 3.1. For the vector field $\mathbf{u} = u_1\,\hat{\mathbf{i}} + u_2\,\hat{\mathbf{j}} + u_3\,\hat{\mathbf{k}}$, we have

$$\nabla \cdot \mathbf{u} = \partial u_1 / \partial x + \partial u_2 / \partial y + \partial u_3 / \partial z. \tag{3.3}$$

Proof. Computing the dot product of two vectors ∇ and **u**, and application of Eq. (1.4.9) yields the result. //

Theorem 3.1. Divergence preserves the linearity and is distributive over linear sum of vectors:

$$\text{div}\,(\mathbf{a} \pm \mathbf{b}) = \text{div}\,\mathbf{a} \pm \text{div}\,\mathbf{b}. \tag{3.4}$$

Proof is simple.

Definition 3.1. A vector filed with vanishing divergence is called *solenoidal.*

Example 3.1. We have

$$\text{div}\,\mathbf{r} = 3. \tag{3.5}$$

Solution. Replacing vector **u** by **r** in Eq. (3.2) and putting from Eqs. (3.1), we easily derive the result. //

Example 3.2. If **a** and **c** are constant vectors, show that

(i) $\text{div}\,(\mathbf{r} \times \mathbf{a}) = 0$, **(ii)** $\text{div}\{(\mathbf{c}.\mathbf{r})\,\mathbf{c}\} = c^2$, **(iii)** $\text{div}\{\mathbf{c} \times (\mathbf{r} \times \mathbf{c})\} = 2c^2$.

Solution. (i) By Eq. (1.4.19), we get

$$\mathbf{r} \times \mathbf{a} = (x, y, z) \times (a_1, a_2, a_3) = \{(a_3 y - a_2 z), (a_1 z - a_3 x), (a_2 x - a_1 y)\}.$$

Taking its divergence, we get

$$\nabla.\,(\mathbf{r} \times \mathbf{a}) = \{\partial/\partial x, \partial/\partial y, \partial/\partial z\}.\{(a_3 y - a_2 z), (a_1 z - a_3 x), (a_2 x - a_1 y)\}$$

$$= (\partial/\partial x)\,(a_3 y - a_2 z) + (\partial/\partial y)\,(a_1 z - a_3 x) + (\partial/\partial z)\,(a_2 x - a_1 y) = 0.$$

(ii) $\quad \mathbf{c}.\,\mathbf{r} = (c_1, c_2, c_3)\,.\,(x, y, z) = c_1 x + c_2 y + c_3 z$

$\Rightarrow$

$$(\mathbf{c}.\,\mathbf{r})\,\mathbf{c} = \{\, c_1 (c_1 x + c_2 y + c_3 z), c_2 (c_1 x + c_2 y + c_3 z),$$

$$c_3 (c_1 x + c_2 y + c_3 z)\,\}.$$

Hence, by Eq. (3.3), its divergence is

$$(\partial/\partial x)\,\{c_1 (c_1 x + c_2 y + c_3 z)\} + (\partial/\partial y)\,\{\, c_2 (c_1 x + c_2 y + c_3 z)\}$$

$$+ (\partial/\partial z)\ \{c_3\,(c_1 x + c_2 y + c_3 z)\} \ = \ c_1^2 + c_2^2 + c_3^2 \ = \ c^2.$$

(iii) By Eq. (1.5.12), we get

$$\mathbf{c} \times (\mathbf{r} \times \mathbf{c}) \ = \ (\mathbf{c}\,.\,\mathbf{c})\,\mathbf{r} \ - (\mathbf{c}\,.\,\mathbf{r})\,\mathbf{c} \ = c^2\,\mathbf{r} - (\mathbf{c}\,.\,\mathbf{r})\,\mathbf{c}.$$

Hence its divergence follows from Example 3.1 and the previous result. //

Example 3.3. Which value of a constant h shall make the vector

$$\mathbf{v} \ = \ (x + 3\,y)\,\hat{\mathbf{i}} + (y - 2\,z)\,\hat{\mathbf{j}} + \ (x + h\,z)\,\hat{\mathbf{k}}$$

solenoidal?

Solution. In view of Eq. (3.3), its divergence is

$$(\partial/\partial x)\,(x + 3\,y) + (\partial/\partial y)\,(y - 2\,z) + (\partial/\partial z)\,(x + h\,z) = 2 + h,$$

which vanishes for $h = -\,2$. //

§ 4. Curl of a vector

Finally, the *cross* product of vector differential operator ∇ and a vector $\mathbf{u}$, yields a vector called the *curl* (or *rot*) of $\mathbf{u}$:

$$\nabla \times \mathbf{u} \ = \{\,\hat{\mathbf{i}}\,(\partial\,/\,\partial x) + \hat{\mathbf{j}}\,(\partial\,/\,\partial y) + \hat{\mathbf{k}}\ (\partial\,/\,\partial z)\} \times \mathbf{u}$$

$$= \ \hat{\mathbf{i}} \times (\partial \mathbf{u}\,/\,\partial x) + \hat{\mathbf{j}} \times (\partial \mathbf{u}\,/\,\partial y) + \hat{\mathbf{k}} \times (\partial \mathbf{u}\,/\,\partial z). \qquad (4.1)$$

If the vector $\mathbf{u}$ has components (u_1, u_2, u_3), for Eq. (1.4.19), above formula reduces to

$$\nabla \times \mathbf{u} \ = \begin{vmatrix} \hat{\mathbf{i}} & \hat{\mathbf{j}} & \hat{\mathbf{k}} \\ \partial/\partial x & \partial/\partial y & \partial/\partial z \\ u_1 & u_2 & u_3 \end{vmatrix}$$

$$= (\partial u_3\,/\partial y - \partial u_2\,/\partial z)\,\hat{\mathbf{i}} + (\partial u_1\,/\partial z - \partial u_3\,/\,\partial x)\,\hat{\mathbf{j}} + (\partial u_2\,/\partial x - \partial u_1/\partial y)\hat{\mathbf{k}}. \quad (4.2)$$

Theorem 4.1. Curl operator also satisfies the linear property:

$$\text{curl}\,(\mathbf{a} \ \pm \mathbf{b}) = \text{curl}\ \mathbf{a} \pm \text{curl}\ \mathbf{b}. \qquad (4.3)$$

Proof is simple.

Definition 4.1. A vector with vanishing curl is called *irrotational*.

Example 4.1. We have

$$\text{curl}\,\mathbf{r} = \mathbf{0}. \tag{4.4}$$

Solution. Replacing the vector **u** by **r** in Eq. (4.1) and putting from Eq. (3.1) we get

$$\text{curl}\,\mathbf{r} = \hat{\mathbf{i}} \times \hat{\mathbf{i}} + \hat{\mathbf{j}} \times \hat{\mathbf{j}} + \hat{\mathbf{k}} \times \hat{\mathbf{k}} = \mathbf{0}, \quad \text{for Eq. (1.4.18). //}$$

Example 4.2. The following vector functions are irrotational:

(i) $$\mathbf{F} = yz\,\hat{\mathbf{i}} + zx\,\hat{\mathbf{j}} + xy\,\hat{\mathbf{k}},$$

(ii) $$\mathbf{F} = \{\sin y + z\cos x,\ \sin z + x\cos y,\ \sin x + y\cos z\},$$

(iii) $$\mathbf{F} = \{x^2 - yz,\ y^2 - zx,\ z^2 - xy\}.$$

Also, show that above functions represent gradient vectors of the following corresponding scalar functions for any arbitrary constant a:

(iv) $$V = xyz + a,$$

(v) $$V = x\sin y + y\sin z + z\sin x + a,$$

(vi) $$V = (1/3)(x^3 + y^3 + z^3) - xyz + a.$$

Solution. (i) In order to establish the irrotational character of the vector we have to check if its curl vanishes. Thus, computing the partial derivatives of the components of **F** and putting for them in Eq. (4.2), vanishing of curl **F** is easily derived.

(ii) $$\partial F_3/\partial y - \partial F_2/\partial z = \cos z - \cos z = 0,$$

$$\partial F_1/\partial z - \partial F_3/\partial x = \cos x - \cos x = 0,$$

$$\partial F_2/\partial x - \partial F_1/\partial y = \cos y - \cos y = 0,$$

$\Rightarrow$

$$\text{curl}\,\mathbf{F} = 0.$$

(iii) $$\partial F_3/\partial y - \partial F_2/\partial z = -x + x = 0,$$

$$\partial F_1 / \partial z - \partial F_3 / \partial x = -y + y = 0, \; \partial F_2 / \partial x - \partial F_1 / \partial y = -z + z = 0,$$

$\Rightarrow$

$$\text{curl } \mathbf{F} = 0.$$

Next, the partial derivatives of V:

$$\partial V / \partial x = y\,z = F_1, \; \partial V / \partial y = z\,x = F_2, \; \partial V / \partial z = x\,y = F_3.$$

Hence, by Eq. (2.3), $\nabla \; V = \mathbf{F}$. Remaining parts can be established similarly. //

§ 5. Some Identities

Theorem 5.1. The vector differential operator ∇ satisfies the following identities:

$$\nabla\,(u\,v) = (\nabla u)\,v + u\,(\nabla v); \text{ i.e. grad } (u\,v) = (\text{grad } u)\,v + u\,(\text{grad } v), \quad (5.1)$$

$$\nabla\,(\mathbf{a}\,.\,\mathbf{b}) = \mathbf{a} \times (\nabla \times \mathbf{b}) + \mathbf{b} \times (\nabla \times \mathbf{a}) + (\mathbf{a}\,.\,\nabla)\,\mathbf{b} + (\mathbf{b}\,.\,\nabla)\,\mathbf{a},$$

i.e.

$$\text{grad } (\mathbf{a}\,.\,\mathbf{b}) = \mathbf{a} \times (\text{curl } \mathbf{b}) + \mathbf{b} \times (\text{curl } \mathbf{a}) + (\mathbf{a}\,.\,\nabla)\,\mathbf{b} + (\mathbf{b}.\,\nabla)\,\mathbf{a}, \quad (5.2)$$

$$\nabla.\,(u\,\mathbf{a}) = (\nabla u).\,\mathbf{a} + u\,(\nabla.\,\mathbf{a}); \text{ i.e. div } (u\,\mathbf{a}) = (\text{grad } u).\,\mathbf{a} + u\,(\text{div } \mathbf{a}), \quad (5.3)$$

$$\nabla \times (u\,\mathbf{a}) = (\nabla u) \times \mathbf{a} + u\,(\nabla \times \mathbf{a}),$$

i.e.

$$\text{curl } (u\,\mathbf{a}) = (\text{grad } u) \times \mathbf{a} + u\,(\text{curl } \mathbf{a}), \quad (5.4)$$

$$\nabla.\,(\mathbf{a} \times \mathbf{b}) = (\nabla \times \mathbf{a})\,.\,\mathbf{b} - \mathbf{a}\,.\,(\nabla \times \mathbf{b}),$$

i.e.

$$\text{div } (\mathbf{a} \times \mathbf{b}) = (\text{curl } \mathbf{a})\,.\,\mathbf{b} - \mathbf{a}\,.\,(\text{curl } \mathbf{b}), \quad (5.5)$$

$$\nabla \times (\mathbf{a} \times \mathbf{b}) = \mathbf{a}\,(\nabla.\,\mathbf{b}) - \mathbf{b}\,(\nabla.\,\mathbf{a}) + (\mathbf{b}\,.\,\nabla)\,\mathbf{a} - (\mathbf{a}\,.\,\nabla)\,\mathbf{b},$$

i.e.

$$\text{curl } (\mathbf{a} \times \mathbf{b}) = \mathbf{a}\,(\text{div } \mathbf{b}) - \mathbf{b}\,(\text{div } \mathbf{a}) + (\mathbf{b}\,.\,\nabla)\,\mathbf{a} - (\mathbf{a}\,.\,\nabla)\,\mathbf{b}. \quad (5.6)$$

Proof. (i) Applying *del* operator on the product function $u\,v$ and using Eq. (2.3), we compute

$$\nabla\,(u\,v) \equiv \{\,\hat{\mathbf{\imath}}\,(\partial / \partial x) + \hat{\mathbf{\jmath}}\,(\partial / \partial y) + \hat{\mathbf{k}}\,(\partial / \partial z)\}\,(u\,v)$$

$$= \hat{\mathbf{\imath}}\,\{\,u\,(\partial v / \partial x) + (\partial u / \partial x)\,v\}$$

$$+ \hat{\mathbf{j}}\, \{ u\,(\partial v / \partial y) + (\partial u / \partial y)\, v\} + \hat{\mathbf{k}}\, \{ u\,(\partial v / \partial z) + (\partial u / \partial z)\, v\}$$

$$= u\, \{\hat{\mathbf{i}}\, (\partial v / \partial x) + \hat{\mathbf{j}}\, (\partial v / \partial y) + \hat{\mathbf{k}}\, (\partial v / \partial z)\}$$

$$+ \{\hat{\mathbf{i}}\, (\partial u/\partial x) + \hat{\mathbf{j}}\, (\partial u/\partial y) + \hat{\mathbf{k}}\, (\partial u/\partial z)\}\, v = u\, (\text{grad } v) + (\text{grad } u)\, v. \, //$$

(ii) By Eq. (4.2),

$$\text{curl } \mathbf{b} = (\partial b_3 / \partial y - \partial b_2 / \partial z)\, \hat{\mathbf{i}} + (\partial b_1 / \partial z - \partial b_3 / \partial x)\, \hat{\mathbf{j}}$$

$$+ (\partial b_2 / \partial x - \partial b_1 / \partial y)\, \hat{\mathbf{k}} = c_1 \hat{\mathbf{i}} + c_2 \hat{\mathbf{j}} + c_3 \hat{\mathbf{k}}, \qquad \text{say.}$$

Using Eq. (1.4.19), we therefore have

$$\mathbf{a} \times (\text{curl } \mathbf{b}) = (a_2 c_3 - a_3 c_2)\, \hat{\mathbf{i}} + (a_3 c_1 - a_1 c_3)\, \hat{\mathbf{j}} + (a_1 c_2 - a_2 c_1)\, \hat{\mathbf{k}}$$

$$= \{a_2 (\partial b_2 / \partial x - \partial b_1 / \partial y) - a_3 (\partial b_1 / \partial z - \partial b_3 / \partial x)\}\, \hat{\mathbf{i}}$$

$$+ \{a_3 (\partial b_3 / \partial y - \partial b_2 / \partial z) - a_1 (\partial b_2 / \partial x - \partial b_1 / \partial y)\}\, \hat{\mathbf{j}}$$

$$+ \{a_1 (\partial b_1 / \partial z - \partial b_3 / \partial x) - a_2 (\partial b_3 / \partial y - \partial b_2 / \partial z)\}\, \hat{\mathbf{k}}. \qquad (5.7)$$

Interchanging vectors **a** and **b** in above identity, we similarly get

$$\mathbf{b} \times (\text{curl } \mathbf{a}) = \{b_2 (\partial a_2 / \partial x - \partial a_1 / \partial y) - b_3 (\partial a_1 / \partial z - \partial a_3 / \partial x)\}\, \hat{\mathbf{i}}$$

$$+ \{b_3 (\partial a_3 / \partial y - \partial a_2 / \partial z) - b_1 (\partial a_2 / \partial x - \partial a_1 / \partial y)\}\, \hat{\mathbf{j}}$$

$$+ \{b_1 (\partial a_1 / \partial z - \partial a_3 / \partial x) - b_2 (\partial a_3 / \partial y - \partial a_2 / \partial z)\}\, \hat{\mathbf{k}}. \qquad (5.8)$$

Also,

$$(\mathbf{a} \,.\, \nabla)\, \mathbf{b} = \{ (a_1, a_2, a_3) \,.\, (\partial/\partial x, \partial/\partial y, \partial/\partial z)\}\, \mathbf{b}$$

$$= (a_1 \partial / \partial x + a_2 \partial / \partial y + a_3 \partial / \partial z)\, (b_1 \hat{\mathbf{i}} + b_2 \hat{\mathbf{j}} + b_3 \hat{\mathbf{k}})$$

$$= \{a_1 (\partial b_1 / \partial x) + a_2 (\partial b_1 / \partial y) + a_3 (\partial b_1 / \partial z)\}\, \hat{\mathbf{i}}$$

$$+ \{a_1 (\partial b_2 / \partial x) + a_2 (\partial b_2 / \partial y) + a_3 (\partial b_2 / \partial z)\}\, \hat{\mathbf{j}}$$

$$+ \{a_1 (\partial b_3 / \partial x) + a_2 (\partial b_3 / \partial y) + a_3 (\partial b_3 / \partial z)\}\, \hat{\mathbf{k}}, \qquad (5.9)$$

and

$$(\mathbf{b}\,.\,\nabla)\,\mathbf{a} = \{b_1(\partial a_1/\partial x) + b_2(\partial a_1/\partial y) + b_3(\partial a_1/\partial z)\}\,\hat{\mathbf{i}}$$

$$+ \{b_1(\partial a_2/\partial x) + b_2(\partial a_2/\partial y) + b_3(\partial a_2/\partial z)\}\,\hat{\mathbf{j}}$$

$$+ \{b_1(\partial a_3/\partial x) + b_2(\partial a_3/\partial y) + b_3(\partial a_3/\partial z)\}\,\hat{\mathbf{k}}\,. \qquad (5.10)$$

Adding all above identities (5.7) - (5.10), and simplifying we get the coefficients of unit vector $\hat{\mathbf{i}}$:

$$\{a_1(\partial b_1/\partial x) + b_1(\partial a_1/\partial x)\} + \{a_2(\partial b_2/\partial x) + b_2(\partial a_2/\partial x)\}$$

$$+ \{a_3(\partial b_3/\partial x) + b_3(\partial a_3/\partial x)\}$$

$$= (\partial/\partial x)(a_1 b_1 + a_2 b_2 + a_3 b_3) = (\partial/\partial x)(\mathbf{a}\,.\,\mathbf{b}).$$

Similarly, the coefficients of unit vectors $\hat{\mathbf{j}}$ and $\hat{\mathbf{k}}$ in above sum are

$$\{a_1(\partial b_1/\partial y) + b_1(\partial a_1/\partial y)\} + \{a_2(\partial b_2/\partial y) + b_2(\partial a_2/\partial y)\}$$

$$+ \{a_3(\partial b_3/\partial y) + b_3(\partial a_3/\partial y)\}$$

$$= (\partial/\partial y)(a_1 b_1 + a_2 b_2 + a_3 b_3) = (\partial/\partial y)(\mathbf{a}\,.\,\mathbf{b});$$

and

$$\{a_1(\partial b_1/\partial z) + b_1(\partial a_1/\partial z)\} + \{a_2(\partial b_2/\partial z) + b_2(\partial a_2/\partial z)\}$$

$$+ \{a_3(\partial b_3/\partial z) + b_3(\partial a_3/\partial z)\}$$

$$= (\partial/\partial z)(a_1 b_1 + a_2 b_2 + a_3 b_3) = (\partial/\partial z)(\mathbf{a}\,.\,\mathbf{b}).$$

Thus, the sum of above identities simplifies to

$$\{(\partial/\partial x)(\mathbf{a}\,.\,\mathbf{b}), (\partial/\partial y)(\mathbf{a}\,.\,\mathbf{b}), (\partial/\partial z)(\mathbf{a}\,.\,\mathbf{b})\} = \text{grad}\,(\mathbf{a}\,.\,\mathbf{b}). \,//$$

(iii) Applying Eq. (3.3) for the vector

$$u\,\mathbf{a} = (u\,a_1, u\,a_2, u\,a_3) \qquad (5.11)$$

we compute

$$\nabla.\,(u\,\mathbf{a}) = \partial(ua_1)/\partial x + \partial(ua_2)/\partial y + \partial(ua_3)/\partial z$$

$$= u\,\{\partial a_1/\partial x + \partial a_2/\partial y + \partial a_3/\partial z\} + \{(\partial u/\partial x)\,a_1 + (\partial u/\partial y)\,a_2 + (\partial u/\partial z)\,a_3\}$$

$$= u\,(\text{div}\,\mathbf{a}) + (a_1, a_2, a_3)\,.\,\{\partial u/\partial x, \partial u/\partial y, \partial u/\partial z\} = \text{RHS of Eq. (5.3).} \,//$$

(iv) Taking the curl of the vector $u\mathbf{a}$ we derive, in view of Eq. (4.2),

$$\text{LHS} = \{\partial (ua_3)/\partial y - \partial (ua_2)/\partial z\}\,\hat{\mathbf{i}} + \{\partial (ua_1)/\partial z - \partial (ua_3)/\partial x\}\,\hat{\mathbf{j}}$$

$$+ \{\partial (ua_2)/\partial x - \partial (ua_1)/\partial y\}\,\hat{\mathbf{k}}$$

$$= \{u\,(\partial a_3/\partial y - \partial a_2/\partial z) + (\partial u/\partial y)\,a_3 - (\partial u/\partial z)\,a_2\}\,\hat{\mathbf{i}}$$

$$+ \{u\,(\partial a_1/\partial z - \partial a_3/\partial x) + (\partial u/\partial z)\,a_1 - (\partial u/\partial x)\,a_3\}\,\hat{\mathbf{j}}$$

$$+ \{u\,(\partial a_2/\partial x - \partial a_1/\partial y) + (\partial u/\partial x)\,a_2 - (\partial u/\partial y)\,a_1\}\,\hat{\mathbf{k}}$$

$$= u\,\{(\partial a_3/\partial y - \partial a_2/\partial z)\,\hat{\mathbf{i}} + (\partial a_1/\partial z - \partial a_3/\partial x)\,\hat{\mathbf{j}} + (\partial a_2/\partial x - \partial a_1/\partial y)\,\hat{\mathbf{k}}\}$$

$$+ \{a_3\,(\partial u/\partial y) - a_2\,(\partial u/\partial z)\}\,\hat{\mathbf{i}} + \{a_1\,(\partial u/\partial z) - a_3\,(\partial u/\partial x)\}\,\hat{\mathbf{j}}$$

$$+ \{a_2\,(\partial u/\partial x) - a_1\,(\partial u/\partial y)\}\,\hat{\mathbf{k}},$$

where the coefficient of u, in view of Eq. (4.2), is curl $\mathbf{a}$, and the remaining vector is $(\nabla u)\times\mathbf{a}$. This establishes the result. //

(v) The cross product of two vectors is given by Eq. (1.4.19). Forming its divergence as per Eq. (3.3), we get

$$\nabla\cdot(\mathbf{a}\times\mathbf{b}) = (\partial/\partial x)\,(a_2 b_3 - a_3 b_2) + (\partial/\partial y)\,(a_3 b_1 - a_1 b_3)$$

$$+ (\partial/\partial z)\,(a_1 b_2 - a_2 b_1). \tag{5.12}$$

On the other hand, using Eq. (4.2), we compute

$$\mathbf{b}.\text{curl }\mathbf{a} = (b_1, b_2, b_3).(\partial a_3/\partial y - \partial a_2/\partial z,\ \partial a_1/\partial z - \partial a_3/\partial x,\ \partial a_2/\partial x - \partial a_1/\partial y)$$

$$= b_1\,(\partial a_3/\partial y - \partial a_2/\partial z) + b_2\,(\partial a_1/\partial z - \partial a_3/\partial x) + b_3\,(\partial a_2/\partial x - \partial a_1/\partial y).$$

Similarly,

$$-\,\mathbf{a}\,.\,\text{curl }\mathbf{b} = a_1\,(-\,\partial b_3/\partial y + \partial b_2/\partial z) + a_2\,(-\,\partial b_1/\partial z + \partial b_3/\partial x)$$

$$+ a_3\,(-\,\partial b_2/\partial x + \partial b_1/\partial y).$$

Adding the last two results and arranging the terms, the RHS simplifies to the one in Eq. (5.12). //

(vi) The cross product of the vectors **a** and **b** is given by Eq. (1.4.19). Forming its curl in view of Eq. (4.2), we derive

$$\text{LHS of Eq. (5.6)} = \{(\partial/\partial y)(a_1 b_2 - a_2 b_1) - (\partial/\partial z)(a_3 b_1 - a_1 b_3)\}\,\hat{\mathbf{i}} +$$

$$\{(\partial/\partial z)(a_2 b_3 - a_3 b_2) - (\partial/\partial x)(a_1 b_2 - a_2 b_1)\}\,\hat{\mathbf{j}} +$$

$$\{(\partial/\partial x)(a_3 b_1 - a_1 b_3) - (\partial/\partial y)(a_2 b_3 - a_3 b_2)\}\,\hat{\mathbf{k}}\,. \qquad (5.13)$$

On the other hand, by Eq. (3.3), we have

$$\mathbf{a}\,(\text{div}\,\mathbf{b}) = (a_1, a_2, a_3)\,\{\partial b_1/\partial x + \partial b_2/\partial y + \partial b_3/\partial z\}$$

$$= \{a_1(\partial b_1/\partial x + \partial b_2/\partial y + \partial b_3/\partial z),\ a_2(\partial b_1/\partial x + \partial b_2/\partial y + \partial b_3/\partial z),$$

$$a_3(\partial b_1/\partial x + \partial b_2/\partial y + \partial b_3/\partial z)\}. \qquad (5.14)$$

Similarly,

$$\mathbf{b}\,(\text{div}\,\mathbf{a}) = \{b_1(\partial a_1/\partial x + \partial a_2/\partial y + \partial a_3/\partial z),$$

$$b_2(\partial a_1/\partial x + \partial a_2/\partial y + \partial a_3/\partial z),\ b_3(\partial a_1/\partial x + \partial a_2/\partial y + \partial a_3/\partial z)\}. \qquad (5.15)$$

Also, **(a.∇) b** and **(b.∇) a** have been evaluated vide Eqs. (5.9) and (5.10). Thus, putting values of all these terms the RHS of the identity (5.6) simplifies to the vector given in Eq. (5.13) establishing the identity. //

Example 5.1. Show that

(i) div grad $r^m = m(m+1)\,r^{m-2}$; **(ii)** curl grad $r^m = \mathbf{0}$;

(iii) curl $\{f(r)\,\mathbf{r}\} = \mathbf{0}$; **(iv)** curl ($u$ grad u) $= \mathbf{0}$.

Solution. (i) grad r^m is computed vide Eq. (2.10). Forming its divergence as per identity (5.3), we get

$$\text{LHS} = m\,\{(\text{grad}\,r^{m-2})\,.\,\mathbf{r} + r^{m-2}\,(\text{div}\,\mathbf{r})\}$$

$$= m\,\{(m-2)\,r^{m-4}\,\mathbf{r}\,.\,\mathbf{r} + 3\,r^{m-2}, \text{ by Eqs. (2.10) and (3.5)}$$

$$= \text{RHS, for } \mathbf{r}\,.\,\mathbf{r} = r^2.$$

(ii) Again, applying curl operator on grad r^m, by Eq. (5.4), we get

LHS $= m\{(\text{grad } r^{m-2}) \times \mathbf{r} + r^{m-2}(\text{curl } \mathbf{r})\} = m(m-2)\, r^{m-4}\, \mathbf{r} \times \mathbf{r} = \mathbf{0}$,

for Eqs. (2.10) and (4.4).

(iii) Again, by identity (5.4),

$$\text{LHS} = \{\text{grad } f(r)\} \times \mathbf{r} + f(r)\,(\text{curl } \mathbf{r}) = (df/dr)\,(\mathbf{r}/r) \times \mathbf{r} = \mathbf{0},$$

by Example 2.3 and Eq. (4.4).

(iv) For identity (5.4),

$$\text{LHS} = (\nabla u) \times (\nabla u) + u\,(\text{curl grad } u) = \mathbf{0} + u\,(\nabla \times \nabla)\, u = \mathbf{0},$$

for Eq. (1.4.18). //

Example 5.2. For constant vectors **a** and **b**, show that

(i) curl $(\mathbf{r} \times \mathbf{a}) = -2\mathbf{a}$, **(ii)** div $\{(\mathbf{r} \times \mathbf{a}) \times \mathbf{b}\} = -2\mathbf{a}\cdot\mathbf{b}$,

(iii) curl $\{(\mathbf{r} \times \mathbf{a}) \times \mathbf{b}\} = \mathbf{b} \times \mathbf{a}$, **(iv)** curl $\{(\mathbf{a}\cdot\mathbf{r})\,\mathbf{b}\} = \mathbf{a} \times \mathbf{b}$,

(v) grad $\{(\mathbf{r} \times \mathbf{a})\cdot(\mathbf{r} \times \mathbf{b})\} = (\mathbf{a} \times \mathbf{r}) \times \mathbf{b} + (\mathbf{b} \times \mathbf{r}) \times \mathbf{a}$.

Solution. (i) Applying the identity (5.6), we get

$$\text{LHS} = \mathbf{r}\,(\nabla\cdot\mathbf{a}) - \mathbf{a}\,(\nabla\cdot\mathbf{r}) + (\mathbf{a}\cdot\nabla)\,\mathbf{r} - (\mathbf{r}\cdot\nabla)\,\mathbf{a}.$$

$= -3\mathbf{a} + \{a_1(\partial/\partial x) + a_2(\partial/\partial y) + a_3(\partial/\partial z)\}\,\mathbf{r}$, for const. **a** and Eq. (3.5)

$$= -3\mathbf{a} + (a_1\hat{\mathbf{i}} + a_2\hat{\mathbf{j}} + a_3\hat{\mathbf{k}}) = -3\mathbf{a} + \mathbf{a} = \text{RHS}.$$

(ii) By Eq. (1.5.12),

$$(\mathbf{r} \times \mathbf{a}) \times \mathbf{b} = (\mathbf{r}\cdot\mathbf{b})\,\mathbf{a} - (\mathbf{a}\cdot\mathbf{b})\,\mathbf{r}. \tag{5.16}$$

Noting the scalar nature of **r. b** and **a. b** being constant, we evaluate its divergence as per identity (5.3):

LHS $= \nabla\cdot\{(\mathbf{r}\cdot\mathbf{b})\,\mathbf{a} - (\mathbf{a}\cdot\mathbf{b})\,\mathbf{r}\} = \{\nabla(\mathbf{r}\cdot\mathbf{b})\}\cdot\mathbf{a} + (\mathbf{r}\cdot\mathbf{b})\,\nabla\cdot\mathbf{a} - (\mathbf{a}\cdot\mathbf{b})\,\nabla\cdot\mathbf{r}$

$$= \{\mathbf{r} \times (\text{curl } \mathbf{b}) + \mathbf{b} \times (\text{curl } \mathbf{r}) + (\mathbf{r}.\ \nabla)\ \mathbf{b} + (\mathbf{b}.\ \nabla)\ \mathbf{r}\}.\ \mathbf{a} - 3\ (\mathbf{a}\ .\ \mathbf{b}),$$

by identity (5.2) and Eq. (3.5). First two terms vanish for invariance of **b** and Eq. (4.4) leaving the expression within curly brackets:

$$(\mathbf{r}.\ \nabla)\ \mathbf{b} + (\mathbf{b}.\ \nabla)\ \mathbf{r} = (x\,\partial / \partial x + y\,\partial / \partial y + z\,\partial / \partial z)\ \mathbf{b}$$

$$+ (b_1\,\partial / \partial x + b_2\,\partial / \partial y + b_3\,\partial / \partial z)\ \mathbf{r} = \mathbf{b}. \qquad (5.17)$$

Thus, the expression reduces to **b. a** – 3 (**a. b**) = – 2 (**a. b**).

(iii) The curl of Eq. (5.16), in view of the identity (5.4), yields:

$$\text{curl } \{(\mathbf{r}.\ \mathbf{b})\ \mathbf{a}\} = \{\nabla\ (\mathbf{r}.\ \mathbf{b})\} \times \mathbf{a} + (\mathbf{r}.\ \mathbf{b})\ (\nabla \times \mathbf{a})$$

$$= \{\ \mathbf{r} \times (\nabla \times \mathbf{b}) + \mathbf{b} \times (\nabla \times \mathbf{r}) + (\mathbf{r}\ .\ \nabla)\ \mathbf{b} + (\mathbf{b}\ .\ \nabla)\ \mathbf{r}\ \} \times \mathbf{a}, \text{ by Eq.(5.2)}$$

$$= \mathbf{b} \times \mathbf{a}, \qquad \text{Eq. (5.17);}$$

and similarly

$$\text{curl } \{\ (\mathbf{a}.\ \mathbf{b})\ \mathbf{r}\ \} = \{\ \nabla\ (\mathbf{a}.\ \mathbf{b})\ \} \times \mathbf{r} + (\mathbf{a}.\ \mathbf{b})\ (\nabla \times \mathbf{r}) = \mathbf{0},$$

for Eq. (4.4) and invariance of **a. b**. Hence, curl of Eq. (5.16) is **b** × **a**.

(iv) In view of the identity (5.4), we have

$$\text{RHS} = \{\nabla\ (\mathbf{a}.\ \mathbf{r})\} \times \mathbf{b} + (\mathbf{a}.\ \mathbf{r})\ (\nabla \times \mathbf{b})$$

$$= \{\mathbf{a} \times (\nabla \times \mathbf{r}) + \mathbf{r} \times (\nabla \times \mathbf{a}) + (\mathbf{a}\ .\nabla)\ \mathbf{r} + (\mathbf{r}\ .\nabla)\ \mathbf{a}\} \times \mathbf{b} = \mathbf{a} \times \mathbf{b},$$

where Eqs. (4.4), (5.2), (5.17) and invariance of **a** are taken into account.

(v) According to Eq. (1.6.1), the scalar product of vectors (**r** × **a**) and (**r** × **b**) is (**r. r**) (**a. b**) – (**r. b**) (**a. r**). Its gradient can be found by Eq. (5.1). Therefore,

$$\text{LHS} = \{\nabla\ (\mathbf{r.r})\}\ (\mathbf{a.b}) + r^2\,\nabla\ (\mathbf{a.b}) - [\{\nabla\ (\mathbf{r.b})\}\ (\mathbf{a.r}) + (\mathbf{r.b})\ \nabla\ (\mathbf{a.r})]$$

$$= 2\mathbf{r}\ (\mathbf{a.b}) - \{\ \mathbf{r} \times (\nabla \times \mathbf{b}) + \mathbf{b} \times (\nabla \times \mathbf{r}) + (\mathbf{r}.\nabla)\ \mathbf{b} + (\mathbf{b}\ .\nabla)\ \mathbf{r}\ \}(\mathbf{a.r})$$

$$- (\mathbf{r}.\ \mathbf{b})\ \{\mathbf{a} \times (\nabla \times \mathbf{r}) + \mathbf{r} \times (\nabla \times \mathbf{a}) + (\mathbf{a}.\nabla)\ \mathbf{r} + (\mathbf{r}.\nabla)\ \mathbf{a}\ \},$$

$$= 2\mathbf{r}\,(\mathbf{a}.\,\mathbf{b}) - \mathbf{b}\,(\mathbf{a}.\,\mathbf{r}) - \mathbf{a}\,(\mathbf{b}.\,\mathbf{r}) = \text{RHS},$$

where Eqs. (2.10), (5.2) and (5.17) are used. //

Example 5.3. If a is **a** constant vector, show that

(i) div $\{(\mathbf{a} \times \mathbf{r})\, r^n\} = \mathbf{0}$, **(ii)** curl $\{(\mathbf{r} \times \mathbf{a}) \times \mathbf{r}\} = 3\,(\mathbf{r} \times \mathbf{a})$,

(iii) curl $\{(\mathbf{a} \times \mathbf{r})\, r^n\} = (n+2)\, r^n\, \mathbf{a} - n\, r^{n-2}\, (\mathbf{a}\,.\,\mathbf{r})\, \mathbf{r}$.

Solution. (i) Applying the identity (5.3), we get

$$\text{LHS} = (\nabla r^n)\,.\,(\mathbf{a} \times \mathbf{r}) + r^n\, \nabla\,.\,(\mathbf{a} \times \mathbf{r})$$

$$= n\, r^{n-2}\, \mathbf{r}\,.\,(\mathbf{a} \times \mathbf{r}) + r^n\, \{(\nabla \times \mathbf{a})\,.\,\mathbf{r} - \mathbf{a}\,.\,(\nabla \times \mathbf{r})\} = \mathbf{0},$$

by Eqs. (1.5.9), (2.10), (4.4) and (5.5).

(ii) Splitting the vector triple product according to Eq. (1.5.12):

$$(\mathbf{r} \times \mathbf{a}) \times \mathbf{r} = (\mathbf{r}\,.\,\mathbf{r})\, \mathbf{a} - (\mathbf{a}\,.\,\mathbf{r})\, \mathbf{r} = r^2 \mathbf{a} - (\mathbf{a}\,.\,\mathbf{r})\, \mathbf{r}.$$

Next, forming the curl of the vector and applying the identity (5.4), we get

$$\text{LHS} = (\nabla r^2) \times \mathbf{a} + r^2 (\nabla \times \mathbf{a}) - \{\nabla\,(\mathbf{a}\,.\,\mathbf{r})\} \times \mathbf{r} - (\mathbf{a}\,.\,\mathbf{r})\,(\nabla \times \mathbf{r})$$

$$= 2\mathbf{r} \times \mathbf{a} - \{\mathbf{a} \times (\text{curl}\,\mathbf{r}) + \mathbf{r} \times (\text{curl}\,\mathbf{a}) + (\mathbf{a}\,.\,\nabla)\,\mathbf{r} + (\mathbf{r}.\,\nabla)\,\mathbf{a}\} \times \mathbf{r}$$

$$= 2\mathbf{r} \times \mathbf{a} - \mathbf{a} \times \mathbf{r} = 3\,(\mathbf{r} \times \mathbf{a}),$$

for Eqs. (2.10), (4.4), (5.2) and invariance of **a**.

(iii) Applying the identity (5.4), we get

$$\text{LHS} = (\nabla r^n) \times (\mathbf{a} \times \mathbf{r}) + r^n\, (\nabla \times (\mathbf{a} \times \mathbf{r})$$

$$= n\, r^{n-2}\, \mathbf{r} \times (\mathbf{a} \times \mathbf{r}) + r^n\, \{\mathbf{a}\,(\nabla.\,\mathbf{r}) - \mathbf{r}\,(\nabla.\,\mathbf{a}) + (\mathbf{r}\,.\,\nabla)\,\mathbf{a} - (\mathbf{a}\,.\,\nabla)\,\mathbf{r}\},$$

by Eqs. (2.10) and (5.6). Splitting the triple product by Eq. (1.5.12), putting from Eqs. (3.5) and noting invariance of the vector **a**, it reduces to

$$n\, r^{n-2}\, \{(\mathbf{r}\,.\,\mathbf{r})\,\mathbf{a} - (\mathbf{r}.\,\mathbf{a})\,\mathbf{r}\} + r^n\, \{3\mathbf{a} - (\mathbf{a}\,.\,\nabla)\,\mathbf{r}\}$$

$$= n\, r^{n-2}\, \{r^2\, \mathbf{a} \;-(\mathbf{r.\,a})\,\mathbf{r}\} + r^n\,(3\mathbf{a}-\mathbf{a}) = \text{RHS, by Eq. (5.17). //}$$

Example 5.4. If $\mathbf{u}$ and f are two point functions the tangential and normal resolutes of $\mathbf{u}$ at any point P on $f(x, y, z) = 0$ are

$$\{(\nabla f)\times(\mathbf{u}\times\nabla f)\}\,/\,(\nabla f)^2 \quad \text{and} \qquad \{\,(\mathbf{u}\,.\nabla f)\,\nabla f\}\,/\,(\nabla f)^2.$$

Solution. Splitting the vector triple product as per Eq. (1.5.12):

$$(\nabla f)\times(\mathbf{u}\times\nabla f) = \{(\nabla f)\,.\,(\nabla f)\}\,\mathbf{u} - \{(\nabla f)\,.\,\mathbf{u}\}\,\nabla f.$$

Dividing by the square of magnitude of the vector ∇f, and arranging the terms, above equation reduces to

$$\mathbf{u} = \{(\nabla f)\times(\mathbf{u}\times\nabla f)\}\,/\,(\nabla f)^2 + \{(\nabla f)\,.\,\mathbf{u}\}\,\nabla f/\,(\nabla f)^2$$

$$= \{(\nabla f)\times(\mathbf{u}\times\nabla f)\}\,/\,(\nabla f)^2 + \{(\nabla f)\,/\,|\nabla f|\}.\,\mathbf{u}\}\;\{\nabla f/\,|\nabla f|\}.$$

Since the gradient vector ∇f is normal to the surface $f = 0$, $(\nabla f)\,/\,|\nabla f|$ gives the unit normal $\overline{\mathbf{n}}$. As such, the second term in above equation represents the normal resolute of vector $\mathbf{u}$ along $\overline{\mathbf{n}}$. Further, the first term containing the cross product with ∇f is evidently perpendicular to ∇f. Hence, it lies in the tangent plane to the surface at P. Thus, the first term represents the tangential resolute of vector $\mathbf{u}$. //

Example 5.5. For a constant vector $\mathbf{c}$, show that

$$\text{curl}\,\{\,\mathbf{c}\times\text{grad}\,(1/r\,)\} + \text{grad}\,\{\mathbf{c}\,.\,\text{grad}\,(1/\,r)\,\} = \mathbf{0}. \qquad (5.18)$$

Solution. By Eq. (2.10),

$$\text{grad}\,(1/\,r\,) = -\,\mathbf{r}\,/\,r^3 \quad (5.19) \quad \Rightarrow \quad \mathbf{c}\times\text{grad}\,(1/r\,) = -\,(r^{-3})\,(\mathbf{c}\times\mathbf{r}).$$

Hence, the first term in Eq. (5.18), for identities (5.4) and (5.6) simplifies to

$$-\,\{\,(\nabla\, r^{-3})\times(\mathbf{c}\times\mathbf{r}) + \,(r^{-3})\,\nabla\times(\mathbf{c}\times\mathbf{r})\,\}$$

$$= 3\,(r^{-5})\,\mathbf{r}\times(\mathbf{c}\times\mathbf{r}) - (r^{-3})\,\{(\nabla.\,\mathbf{r})\,\mathbf{c} - (\nabla.\,\mathbf{c})\,\mathbf{r} + (\mathbf{r}\,.\,\nabla)\,\mathbf{c} - (\mathbf{c}\,.\,\nabla)\,\mathbf{r}\}$$

$$= 3\,(r^{-5})\{(\mathbf{r}\,.\,\mathbf{r})\,\mathbf{c} - (\mathbf{r}\,.\,\mathbf{c})\,\mathbf{r}\} - (r^{-3})\,(3\mathbf{c}-\mathbf{c})$$

$$= (\mathbf{c}\,/\,r^3) - 3\,(\mathbf{r}\,.\,\mathbf{c})\,\mathbf{r}\,/r^5,$$

for Eqs. (1.5.12), (2.10), (5.17) and invariance of **c**.

Similarly, **c**. grad $(1/r) = -(r^{-3})$ (**c** . **r**), and its gradient for Eqs. (2.10), (5.1) and (5.2) is

$$-\{(\nabla r^{-3})(\mathbf{c}\cdot\mathbf{r})\} - (r^{-3})\nabla(\mathbf{c}\cdot\mathbf{r})$$

$$= 3(\mathbf{r}.\mathbf{c})\,\mathbf{r}/r^5 - (r^{-3})\{\mathbf{c}\times(\text{curl }\mathbf{r}) + \mathbf{r}\times(\text{curl }\mathbf{c}) + (\mathbf{c}.\nabla)\,\mathbf{r} + (\mathbf{r}.\nabla)\,\mathbf{c}\}$$

$$= 3(\mathbf{r}\cdot\mathbf{c})\,\mathbf{r}/r^5 - \mathbf{c}/r^3, \text{ by Eqs. (4.4), (5.17).}$$

Hence, their sum becomes a null vector. //

§ 6. Second order derivation of vectors

6.1. *Laplace* equation: The second order derivative

$$\nabla^2 = \nabla\cdot\nabla = \partial^2/\partial x^2 + \partial^2/\partial y^2 + \partial^2/\partial z^2 \tag{6.1}$$

defines the *Laplacian* operator. The function satisfying $\nabla^2 f = 0$ is called *harmonic* and the result is called the *Laplace* equation.

Theorem 6.1. Various combinations of three types of operations defined in preceding sections satisfy

$$\text{div (grad } f) \equiv \nabla.\,\nabla f = \nabla^2 f = \partial^2 f/\partial x^2 + \partial^2/\partial y^2 + \partial^2/\partial z^2, \tag{6.2}$$

$$\text{div (curl } \mathbf{a}) \equiv \nabla\cdot\nabla\times\mathbf{a} = 0, \tag{6.3}$$

$$\text{curl (grad } f) \equiv \nabla\times(\nabla f) = \mathbf{0}, \tag{6.4}$$

$$\text{curl (curl } \mathbf{a}) \equiv \nabla\times(\nabla\times\mathbf{a}) = \nabla(\nabla.\,\mathbf{a}) - \nabla^2\mathbf{a}. \tag{6.5}$$

Proof. **(i)** The first result follows immediately from Eqs. (2.3) and (3.2).

(ii) Next, evaluating curl **a** by Eq. (4.2):

$$\text{curl } \mathbf{a} = \{(\partial a_3/\partial y - \partial a_2/\partial z), (\partial a_1/\partial z - \partial a_3/\partial x), (\partial a_2/\partial x - \partial a_1/\partial y)\}, \tag{6.6}$$

and computing its divergence as per Eq. (3.3), we get

$$\nabla\cdot\nabla\times\mathbf{a} = (\partial/\partial x)(\partial a_3/\partial y - \partial a_2/\partial z)$$

$$+ (\partial / \partial y)(\partial a_1 / \partial z - \partial a_3 / \partial x) + (\partial / \partial z)(\partial a_2 / \partial x - \partial a_1 / \partial y) = 0.$$

(iii) The gradient vector is given by Eq. (2.3). Its curl may be evaluated from Eq. (4.2):

$$\nabla \times (\nabla f) = \{ (\partial^2 f / \partial y\, \partial z - \partial^2 f / \partial z\, \partial y), (\partial^2 f / \partial z\, \partial x - \partial^2 f / \partial x\, \partial z),$$

$$(\partial^2 f / \partial x\, \partial y - \partial^2 f / \partial y\, \partial x) \} = \mathbf{0}.$$

(iv) Curl **a** is evaluated in Eq. (6.6): curl **a** = $\{c_1, c_2, c_3\}$, where the vector **c** is given by Eq. (6.6). Again, forming its curl vide Eq. (6.6), we derive

$$\text{curl (curl } \mathbf{a}) = \{(\partial c_3 / \partial y - \partial c_2 / \partial z), (\partial c_1 / \partial z - \partial c_3 / \partial x), (\partial c_2 / \partial x - \partial c_1 / \partial y)\}. \tag{6.7}$$

Thus, putting for **c** from Eq. (6.6), we evaluate the components in above vector:

$$\partial c_3 / \partial y - \partial c_2 / \partial z = (\partial / \partial y)(\partial a_2 / \partial x - \partial a_1 / \partial y) - (\partial / \partial z)(\partial a_1 / \partial z - \partial a_3 / \partial x)$$

$$= \partial^2 a_2 / \partial x\, \partial y - \partial^2 a_1 / \partial y^2 - \partial^2 a_1 / \partial z^2 + \partial^2 a_3 / \partial x\, \partial z$$

$$= (\partial / \partial x)\{\partial a_1 / \partial x + \partial a_2 / \partial y + \partial a_3 / \partial z\} - \{\partial^2 a_1 / \partial x^2 + \partial^2 a_1 / \partial y^2 - \partial^2 a_1 / \partial z^2\}$$

$$= (\partial / \partial x)(\nabla . \mathbf{a}) - \nabla^2 a_1.$$

Similarly, the remaining components are

$$(\partial c_1 / \partial z - \partial c_3 / \partial x) = (\partial / \partial y)(\nabla . \mathbf{a}) - \nabla^2 a_2,$$

and

$$\partial c_2 / \partial x - \partial c_1 / \partial y = (\partial / \partial z)(\nabla . \mathbf{a}) - \nabla^2 a_3.$$

Hence, the RHS of Eq. (6.7) is

$$\{(\partial / \partial x)(\nabla . \mathbf{a}) - \nabla^2 a_1, (\partial / \partial y)(\nabla . \mathbf{a}) - \nabla^2 a_2, (\partial / \partial z)(\nabla . \mathbf{a}) - \nabla^2 a_3\}$$

$$= \nabla (\nabla . \mathbf{a}) - \nabla^2 \mathbf{a}. //$$

Theorem 6.2. The gradient of a harmonic function is both solenoidal and irrotational vector field.

Proof. Since $\nabla \,.\, (\nabla f) = \nabla^2 f = 0$, as per hypothesis, makes the vector solenoidal. Further, the irrotational character of the vector field ∇f is already established vide Eq. (6.4). //

Example 6.1. The function $f = 1/r$ is harmonic, where $r = |\,\mathbf{r}\,|$ is the length of the position vector **r**.

CHAPTER 3

VECTORS IN ELECTRIC AND MAGNETIC FIELDS

§ 1. Electric field

An electric field surrounds an electric charge exerting forces: attracting or repelling on other charges in the field. It is often abbreviated as *E*-field. Mathematically speaking it is a vector field that associates to each point in space the (electrostatic or Coulomb) force per unit of charge exerted on an infinitesimal positive test charge at rest at that point. The SI [1] unit for electric field strength is volt per metre (V/m). However, Newtons per coulomb (N/C) is also used as an alternate unit of electric field strength. Electric fields are created by electric charges, or by time-varying magnetic fields. These fields are important in physics, and are exploited practically in electrical technology. In terms of atoms, the electric field is responsible for the attractive force between the atomic nucleus and electrons holding the atom together, and the forces between atoms causing chemical bonding. Electric fields and magnetic fields are both manifestations of the electromagnetic force - one of the four fundamental forces (or interactions) of nature.

Definition. Let two particles located at positions x_o and x_1 be charged with electric charges q_o and q_1 respectively. By Coulomb's law the latter particle exerts a force

$$\mathbf{F} = \frac{1}{4\pi\varepsilon_o} \cdot \frac{q_o\, q_1}{(x_1 - x_o)^2} \cdot \hat{\mathbf{r}}_1, \tag{1.1}$$

on the former particle, where $\hat{\mathbf{r}}_1$ is the unit vector in the direction from point x_1 to point x_o, and ε_o is the electric constant (also known as "the absolute permittivity of free space") in $C^2\, m^{-2}\, N^{-1}$.

If the charges q_o and q_1 are of the same sign, above force is positive and is directed away from the other charge, i.e. the particles repel each other. On contrary, if the charges are of opposite signs the force is negative and the particles attract each other. For convenience, the Coulomb force on the charge q_o, the expression in Eq. (1.1) may be divided by q_o

---

[1] SI = système internationale (French) = international system (of units).

and expression depends only on the charge q_1 (called the *source* charge).

$$\mathbf{E}\,(x_o) = \mathbf{F} / q_o = \frac{1}{4\pi\varepsilon_o} \cdot \frac{q_1}{(x_1 - x_o)^2} \cdot \hat{\mathbf{r}}_1, \qquad (1.2)$$

giving the *electric field* at point x_o due to the point charge q_1. It is a vector quantity measuring the Coulomb force per unit charge exerting on the positive point charge at x_o. This (electric) field becomes infinite at the location of the (*source*) charge itself. From Eq. (1.2), it is clear that the electric field due to a point charge is everywhere directed away from the charge if it is positive, and toward the charge if it is negative, and its magnitude decreases with the inverse square of the distance from the charge.

1.1. Multiple charges

In case of multiple charges, the resultant Coulomb force on a charge is computed by adding the vector (forces) due to each charge. Thus, the electric field satisfies the

***Superposition principle*:** the total electric field at a point due to multiple charges is just equal to the vector sum of the electric fields at that point due to individual charges.

$$\mathbf{E}\,(x) = \mathbf{E}_1\,(x) + \mathbf{E}_2\,(x) + \ldots + \mathbf{E}_n\,(x) = \frac{1}{4\pi\varepsilon_o} \cdot \sum_{i=1}^{n} \left\{ \frac{q_i}{(x_i - x_o)^2} \right\} \hat{\mathbf{r}}_i\,, \qquad (1.3)$$

where $\hat{\mathbf{r}}_i$ is the unit vector in the direction from point x_i to point x_o.

Note 1.1. The Coulomb force on a charge of magnitude q at any point in space is equal to the product of the charge and the electric field at that point $\mathbf{F} = q\,\mathbf{E}$.

1.2. Continuous distribution of charge

Let the space be charged with a continuous distribution of charge of density $\rho\,(x)$ coulombs per cubic meter, the electric field due to such is obtained by considering the charge $\rho\,(x')\,dV$ in each infinitesimal volume dV of space at point x' as a point charge. Its electric field $d\,\mathbf{E}\,(x)$ at a point x is computed by:

$$d\,\mathbf{E}\,(x) = \frac{1}{4\pi\varepsilon_o} \cdot \frac{\rho\,(x')\,d\,V}{(x' - x_o)^2} \cdot \hat{\mathbf{r}}', \tag{1.4}$$

where $\hat{\mathbf{r}}'$ is the unit vector pointing from x' to x, then adding up the contributions from all the increments of volume by integrating over the volume of the charge distribution V:

$$\mathbf{E}\,(x) = \frac{1}{4\pi\varepsilon_o} \cdot \iiint_V \frac{\rho\,(x')\,d\,V}{(x' - x_o)^2} \cdot \hat{\mathbf{r}}'. \tag{1.5}$$

§ 2. Magnetic field

The magnetic field at a point in a plane is a vector quantity having a specific direction associated with the field and the field strength. Cf. Fig. 1.1.1. The direction of the arrow determines the direction of the magnetic field, while the length of the arrow gives the strength of the field. Let this vector be resolved along two mutually perpendicular Cartesian coordinate axes Ox and Oy. As explained in § 3 of Chapter1, let $\mathbf{r}_x$ and $\mathbf{r}_y$ be two components of the field, called the x- and y-components. By Pythagoras theorem, the strength of the magnetic field is measured in terms of these components:

$$|\,\mathbf{r}\,| = \sqrt{(\mathbf{r}_x{}^2 + \mathbf{r}_y{}^2)}. \tag{2.1}$$

Considering the field in 3-dimensional space, there exist three mutually perpendicular components of the magnetic field, say $\mathbf{r}_x$, $\mathbf{r}_y$ and $\mathbf{r}_z$. The strength of the magnetic field is then measured by the formula

$$|\,\mathbf{r}\,| = \sqrt{(\mathbf{r}_x{}^2 + \mathbf{r}_y{}^2 + \mathbf{r}_z{}^2)}, \tag{2.2}$$

which is the exact generalization of the Eq. (2.1).

CHAPTER 4

TENSORS (CARTESIAN)

§ 1. Coordinate system in V_n

The concept of orthogonality of vectors (whether in Euclidean or Non-Euclidean spaces) cannot be extended beyond three vectors. Therefore, in a space of dimension higher than three it is replaced by the concept of linear independence of vectors.

Let V_n be an n-dimensional vector space equipped with n linearly independent variables x^i 's (i = 1, 2,..., n). If n linearly independent vectors $\hat{\mathbf{e}}_i$'s form a basis of the space with respect to above variables, the set

$$\{ x^i \} \equiv (x^1, x^2, ..., x^n) \qquad (1.1)$$

form a coordinate system in V_n. For brevity, the coordinates of a point $P \in V_n$ with respect to some origin O and above basis are denoted by (x^i).

We now consider a non-singular linear transformation of V_n onto itself given by

$$\eta : (x^i) \to (\bar{x}^a). \qquad (1.2)$$

As such, there also exist its inverse transformation:

$$\eta^{-1} : (\bar{x}^a) \to (x^i). \qquad (1.3)$$

Because of linear independence of the coordinates, there follow the relations

$$\delta^i_j \equiv \partial x^i / \partial x^j = (\partial x^i / \partial \bar{x}^a)(\partial \bar{x}^a / \partial x^j), \qquad (1.4)$$

and

$$\bar{\delta}^a_b \equiv \partial \bar{x}^a / \partial \bar{x}^b = (\partial \bar{x}^a / \partial x^i)(\partial x^i / \partial \bar{x}^b), \qquad (1.5)$$

from Taylor's theorem. The symbols δ^i_j, called *Kronecker deltas* (after a German mathematician Leopold Kronecker (1823-91 A.D.), are given by

$$\delta^i_j = \begin{cases} 1 & \text{when } i = j, \\ 0 & \text{when } i \neq j. \end{cases} \qquad (1.6)$$

Theorem 1.1. For any functions $X^j(x^k)$ there holds

$$\delta'_j X^j = X^i. \tag{1.7}$$

Proof. Following Einstein's summation convention the left member of Eq. (1.7) denotes the sum

$$\delta'_1 X^1 + \delta'_2 X^2 + ... + \delta'_n X^n,$$

which, for Eq. (1.6), reduces to X^i. //

The classical approach defining scalars (having magnitude only) and vectors (with magnitude as well as direction) is just unable to meet the requirements of a space of dimension higher than three. In addition to the axiomatic approach (which is the latest) for such quantities, we present a categorization of various geometric entities according to their different coordinate transformation laws vide Eq. (1.2) in the subsequent sections of this chapter.

§ 2. Scalars

Definition 2.1. A function $f(x^1, x^2,..., x^n)$, briefly denoted as $f(x^i)$, is called a *scalar* or *invariant* if it remains unaltered under the transformation vide Eq. (1.2):

$$f(x^i) = f(\bar{x}^a). \tag{2.1}$$

In other words, the transformation vide Eq. (1.2) maps the scalar function $f(x^i)$ onto itself.

Example 2.1. Distance between two points is a scalar function.

Example 2.2. Magnitude of a vector is a scalar function.

Example 2.3. Angle between two directions is a scalar function.

Note 2.1. Proofs of these examples is furnished in [12], Chapter 15.

Note 2.2. The pressure and density of a fluid and the electrostatic potential are other examples of scalars.

Scalars can be variable functions or constants as well.

§ 3. Einstein's summation convention

When an index is repeated in a term it is called '*dummy*' or '*umbral*'; else a '*free*' index. Following Einstein's summation convention a dummy index is ought to be summed over its range even without writing the summation symbol $\sum$. Thus, the sum of n terms:

$$x_1\,\alpha_1 + x_2\,\alpha_2 + \ldots + x_n\,\alpha_n = \sum x_i\,\alpha_i, \tag{3.1}$$

can be simply denoted by $x_i\,\alpha_i$, where the index i runs from 1 to n. Similarly, the sum

$$x_1\,\hat{\mathbf{e}}_1 \oplus x_2\,\hat{\mathbf{e}}_2 \oplus x_3\,\hat{\mathbf{e}}_3 = \sum x_i\,\hat{\mathbf{e}}_i, \tag{3.2}$$

could be taken as $x_i\,\hat{\mathbf{e}}_i$ only for the range $i = 1, 2, 3$.

In the subsequent chapters, we deal with an n-dimensional space with Cartesian coordinates $x^1, x^2, \ldots x^n$; briefly denoted by (x^i). As such, the Latin indices a to m as well as Greek indices α, β, γ assume positive integral values from 1 to n unless a contrary is stated.

In the following, we mention some properties of dummy indices.

Note 3.1. A dummy index can be replaced by any other dummy index ranging over the same numbers.

Thus, the linear combination in Eq. (3.1) can also be denoted by $x_j\,\alpha_j$ where j also runs from 1 to n; for either of $x_i\,\alpha_i$ and $x_j\,\alpha_j$ give the same linear combination.

Note 3.2. The dummy indices (having same range) in a term are also interchangeable:

$$g_{ij}\,dx^i\,dx^j = g_{ji}\,dx^j\,dx^i,$$

as either side gives the sum of same n^2 terms (cf. [12], Eq. (15.2.1)).

Note 3.3. To avoid confusion a dummy index should not occur more than twice in a single term.

§ 4. Contravariant vector

Definition 4.1. A set of functions $X^i(x^j)$ mapping onto $\overline{X}^a(\overline{x}^b)$

under the transformations vide Eq. (1.2) as per

$$X^i = \bar{x}^a \,(\partial x^i / \partial \bar{x}^a), \tag{4.1a}$$

or, equivalently,

$$X^i \,(\partial \bar{x}^a / \partial x^i) = \bar{x}^a, \tag{4.1b}$$

is said to form the components of a *contravariant* vector.

Example 4.1. The tangent vector to a curve is a contravariant vector.

Solution. Let C: $x^i = x^i(s)$ be a curve in V_n with two neighbouring points P (x^i) and Q $(x^i + dx^i)$. The limiting position of the chord $\overrightarrow{PQ}$ with components dx^i determines the tangent vector to C at P. Applying the Taylor's formula for x^i treated as a function of $\bar{x}^a$'s, we obtain

$$dx^i = (\partial x^i / \partial \bar{x}^a)\, d\bar{x}^a, \tag{4.2}$$

where $d\bar{x}^a$ form the components of the tangent vector to the deformed curve $\bar{C}$ at $\bar{P}$. We note that the transformations vide Eqs. (4.2) of dx^i are in accordance with Eqs. (4.1a), so we have the statement. //

§ 5. Covariant vector

Definition 5.1. A set of functions $Y_i(x^j)$ mapping onto $\bar{Y}_a(\bar{x}^a)$ under the transformations vide Eqs. (1.2) according to

$$Y_i = \bar{Y}_a (\partial \bar{x}^a / \partial x^i), \tag{5.1a}$$

or, equivalently,

$$Y_i (\partial x^i / \partial \bar{x}^a) = \bar{Y}_a, \tag{5.1b}$$

is said to form the components of a *covariant* vector.

Example 5.1. The normal to a surface

$$f(x^i) = c \text{ (const.)} \tag{5.2}$$

is a covariant vector.

Solution. The functional relation in Eq. (5.2) loses the independence of all the n variables x^i's; out of which, there exist at most $n - 1$ independent variables. Hence, the surface is of dimension $n - 1$. Further,

$f(x^i)$ being scalar functions satisfy Eq. (2.1):

$$f(x^i) = f(\bar{x}^a) = c \quad \Rightarrow \quad df(x^i) = df(\bar{x}^a) = 0;$$

or,

$$(\partial f / \partial x^i)\, dx^i = (\partial f / \partial \bar{x}^a)\, d\bar{x}^a = 0. \tag{5.3}$$

dx^i and $d\bar{x}^a$ being components of the tangent vectors, Eqs. (5.3) imply the normal character of $\partial f / \partial x^i$ and $\partial f / \partial \bar{x}^a$. For Eq. (4.2), the relations (5.3) can be rewritten as

$$\{(\partial f / \partial x^i) - (\partial f / \partial \bar{x}^a)(\partial \bar{x}^a / \partial x^i)\}\, dx^i = 0,$$

or,

$$\partial f / \partial x^i = (\partial f / \partial \bar{x}^a)\,(\partial \bar{x}^a / \partial x^i), \tag{5.4}$$

as dx^i 's are linearly independent. The transformations vide Eqs. (5.4) for the components of the normal vector are in agreement with Eq. (5.1a). Hence, we have the statement. //

§ 6. Contravariant tensor of order two

Definition 6.1. A set of n^2 functions $T^{ij}(x^k)$ of V_n mapping onto $\bar{T}^{ab}(\bar{x}^c)$ under the transformations vide Eq. (1.2) as per the rule

$$T^{ij} = \bar{T}^{ab}\,(\partial x^i / \partial \bar{x}^a)\,(\partial x^j / \partial \bar{x}^b) \tag{6.1}$$

is said to form the components of a *second order* (or *bivalent*) *contravariant tensor*.

Following L.P. Eisenhart [1], we define the rank of this tensor.

Definition 6.2. The rank of the matrix $((T^{ij}))$ is called the *rank* of the tensor.

Example 6.1. The outer product of two contravariant vectors is a second order contravariant tensor.

Solution. Let $A^i(x^j)$ and $B^k(x^h)$ form the components of two contravariant vectors satisfying the coordinate transformation law vide Eq. (4.1a):

$$A^i = \bar{A}^a\,(\partial x^i / \partial \bar{x}^a), \qquad B^k = \bar{B}^b\,(\partial x^k / \partial \bar{x}^b).$$

Their multiplication yields

$$A^i B^k = (\bar{A}^a \bar{B}^b)(\partial x^i / \partial \bar{x}^a)(\partial x^k / \partial \bar{x}^b). \tag{6.2}$$

Putting

$$T^{ik} \equiv A^i B^k, \qquad \bar{T}^{ab} \equiv A^a \bar{B}^b,$$

and comparing Eq. (6.2) with Eq. (6.1), we get the statement. //

Example 6.2. The tensor in previous example is of rank one.

Solution. The components of the tensor form the determinant

$$|A^i B^j| = \begin{vmatrix} A^1 B^1 & A^1 B^2 & \dots & A^1 B^n \\ A^2 B^1 & A^2 B^2 & \dots & A^2 B^n \\ . & . & \dots & . \\ A^n B^1 & A^n B^2 & \dots & A^n B^n \end{vmatrix}.$$

Taking out the common multiples $B^1, B^2, \dots, B^n$ of the respective columns the determinant can be seen zero. Besides, every minor of any order from 2 to $n-1$ also vanishes. Hence, the rank of the determinant is one. //

§ 7. Contravariant tensor of rth order

Definition 7.1. A set of n^r functions $T^{i_1 i_2 \dots i_r}(x^j)$ of V_n mapping onto $\bar{T}^{a_1 a_2 \dots a_r}(\bar{x}^b)$ under the transformations in Eq. (1.2) as per

$$T^{i_1 i_2 \dots i_r} = \bar{T}^{a_1 a_2 \dots a_r}(\partial x^{i_1} / \partial \bar{x}^{a_1})(\partial x^{i_2} / \partial \bar{x}^{a_2}) \dots (\partial x^{i_r} / \partial \bar{x}^{a_r}) \tag{7.1}$$

is said to form the components of an rth order *contravariant tensor.*

Example 7.1. The outer product of r contravariant vectors gives an rth order contravariant tensor.

Solution. Proof is similar to that in Example 6.1. //

§ 8. Covariant tensor of order two

Definition 8.1. n^2 functions $S_{ij}(x^k)$ of V_n mapping onto $\bar{S}_{ab}(\bar{x}^c)$ under the transformations in Eq. (1.2) as per

$$S_{ij} = \overline{S}_{ab}\,(\partial \overline{x}^a / \partial x^i)\,(\partial \overline{x}^b / \partial x^j) \tag{8.1}$$

are said to form the components of a *second order* (or *bivalent*) *covariant tensor*.

Example 8.1. The outer product of two covariant vectors is a second order covariant tensor.

Solution. Let $A_i\,(x^j)$ and $B_k\,(x^h)$ be components of two covariant vectors satisfying the coordinate transformation law vide Eq. (5.1a):

$$A_i = \overline{A}_a\,(\partial \overline{x}^a / \partial x^i), \qquad B_k = \overline{B}_b\,(\partial \overline{x}^b / \partial x^k).$$

Their multiplication yields

$$A_i\,B_k = (\overline{A}_a\,\overline{B}_b)\,(\partial \overline{x}^a / \partial x^i)\,(\partial \overline{x}^b / \partial x^k), \tag{8.2}$$

which, on comparison with Eq. (8.1), establishes the statement. //

Example 8.2. If $A_{ij}X^iX^j$ is an invariant for an arbitrary contravariant vector X^i, then $A_{ij} + A_{ji}$ span a covariant tensor of second order. Further,

$$A_{ij}X^iX^j = 0 \tag{8.3}$$

$\Rightarrow$

$$A_{ij} + A_{ji} = 0. \tag{8.4}$$

Solution. As per hypothesis, we have

$$A_{ij}X^iX^j = \overline{A}_{ab}\,\overline{X}^a\,\overline{X}^b$$

from Eq. (2.1). Interchange of dummy indices in the pairs i, j and a, b in above equation also gives

$$A_{ji}X^jX^i = \overline{A}_{ba}\,\overline{X}^b\,\overline{X}^a.$$

Adding the two relations there follows

$$(A_{ij} + A_{ji})\,X^iX^j = (\overline{A}_{ab} + \overline{A}_{ba})\,\overline{X}^a\,\overline{X}^b,$$

or, for Eq. (4.1a),

$$\{\,(A_{ij} + A_{ji})\,(\partial x^i / \partial \overline{x}^a)\,(\partial x^j / \partial \overline{x}^b) - (\overline{A}_{ab} + \overline{A}_{ba})\,\}\,\overline{X}^a\,\overline{X}^b = 0.$$

For $\overline{X}^{a}$ being linearly independent components of an arbitrary vector, above equation concludes

$$(A_{ij}+A_{ji})\,(\partial x^{i}/\partial \overline{x}^{a})\,(\partial x^{j}/\partial \overline{x}^{b}) = \overline{A}_{ab}+\overline{A}_{ba},$$

or, equivalently,

$$A_{ij}+A_{ji} = (\overline{A}_{ab}+\overline{A}_{ba})\,(\partial \overline{x}^{a}/\partial x^{i})\,(\partial \overline{x}^{b}/\partial x^{j}). \qquad (8.5)$$

A comparison of Eq. (8.5) with Eq. (8.1) establishes the tensorial character of the functions $A_{ij}+A_{ji}$. Next, an interchange of i, j in Eq. (8.3) gives

$$A_{ji}X^{j}X^{i} = 0,$$

which, on addition to Eq. (8.3), determines

$$(A_{ij}+A_{ji})\,X^{i}X^{j} = 0. \qquad (8.6)$$

Again, for X^{i} being arbitrary, Eq. (8.6) concludes Eq. (8.4). //

§ 9. Covariant tensor of sth order

Definition 9.1. n^{s} functions $S_{j_1 j_2 \cdots j_s}(x^{k})$ of V_n mapping onto $\overline{S}_{b_1 b_2 \cdots b_s}(\overline{x}^{c})$ under the transformations vide Eq. (1.2) as per

$$\overline{S}_{j_1 j_2 \ldots j_s} = \overline{S}_{b_1 b_2 \cdots b_s}(\partial \overline{x}^{b_1}/\partial x^{j_1})(\partial \overline{x}^{b_2}/\partial x^{j_2})\ldots(\partial \overline{x}^{b_s}/\partial x^{j_s}) \qquad (9.1)$$

are said to form the components of an sth *order covariant tensor*.

Example 9.1. The outer product of s covariant vectors is an sth order covariant tensor.

Solution. Proof is similar to that in Example 8.1. //

§ 10. Mixed tensor of type (1, 1)

Definition 10.1. n^{2} functions $T^{i}{}_{j}(x^{k})$ of V_n mapping onto $\overline{T}_{b}^{a}(\overline{x}^{c})$ under the transformations in Eq. (1.2) according to

$$T^i_j = \overline{T}^a_b (\partial x^i / \partial \overline{x}^a)(\partial \overline{x}^b / \partial x^j) \tag{10.1}$$

are said to form the components of a *mixed tensor of the type* (1, 1).

It may be noted that such a tensor has one contravariant valency and one covariant valency. Therefore, it is also a second order (or bivalent) tensor.

Definition 10.2. Sum of the valencies is called the *order* [2)] of the tensor.

Example 10.1. The outer product of a contravariant vector and a covariant vector is a mixed tensor of type (1, 1).

Solution. Let $X^i(x^j)$ and $Y_k(x^h)$ form the components of contravariant and covariant vectors respectively. Their coordinate transformation laws are given by Eqs. (4.1a) and (5.1) respectively. Multiplying these equations, we get

$$X^i Y_k = (\overline{X}^a \ \overline{Y}_b)(\partial x^i / \partial \overline{x}^a)(\partial \overline{x}^b / \partial x^k). \tag{10.2}$$

Comparing Eq. (10.2) with Eq. (10.1), we get the statement. //

Example 10.2. Kronecker deltas form the components of a mixed tensor of type (1, 1).

Solution. Analogous to Eqs. (1.4) and (1.5), there also results

$$\partial \overline{x}^a / \partial x^j = (\partial \overline{x}^a / \partial \overline{x}^b)(\partial \overline{x}^b / \partial x^j), \tag{10.3}$$

from Eq. (1.2). Consequently, Eq. (1.4) reduces to

$$\delta^i_j = (\partial x^i / \partial \overline{x}^a)(\partial \overline{x}^a / \partial \overline{x}^b)(\partial \overline{x}^b / \partial x^j)$$

$$= \overline{\delta}^a_b (\partial x^i / \partial \overline{x}^a)(\partial \overline{x}^b / \partial x^j), \tag{10.4}$$

by Eq. (1.5). Being in agreement with Eq. (10.1), this proves the statement. //

---

2) M.M. Lipschutz [3] calls it *rank*.

§ 11. Mixed tensor of type (*r*, *s*)

Definition 11.1. $n^{(r+s)}$ functions $T^{i_1 i_2 \dots i_r}_{j_1 j_2 \dots j_s}(x^k)$ of V_n mapping onto

$\overline{T}^{a_1 a_2 \dots a_r}_{b_1 b_2 \dots b_s}(\overline{x}^c)$ under transformation in Eq. (1.2) according to

$$T^{i_1 i_2 \dots i_r}_{j_1 j_2 \dots j_s} = \overline{T}^{a_1 a_2 \dots a_r}_{b_1 b_2 \dots b_s}(\partial x^{i_1}/\partial \overline{x}^{a_1})(\partial x^{i_2}/\partial \overline{x}^{a_2})\dots(\partial x^{i_r}/\partial \overline{x}^{a_r}).$$

$$.(\partial \overline{x}^{b_1}/\partial x^{j_1})(\partial \overline{x}^{b_2}/\partial x^{j_2})\dots(\partial \overline{x}^{b_s}/\partial x^{j_s}) \qquad (11.1)$$

are said to form the components of a *mixed tensor of type* (*r*, *s*). The sum $r+s$ determines the *order* of the tensor.

Theorem 11.1. The coordinate transformation laws of vectors and tensors possess group property.

Proof. Let the coordinate system $(\overline{x}^a)$ be further transformed under some non-singular linear transformations

$$(\overline{x}^a) \to (\overline{\overline{x}}^\alpha), \qquad (11.2)$$

where $\alpha = 1, 2, \dots, n$. Thus, the composition of two non-singular transformations vide Eqs. (1.2) and (11.2) gives a direct transformation of (x^i) onto $(\overline{\overline{x}}^\alpha)$:

$$(x^i) \to (\overline{\overline{x}}^\alpha). \qquad (11.3)$$

Hence, by Taylor's theorem, there results

$$\partial x^i/\partial \overline{\overline{x}}^\alpha = (\partial x^i/\partial \overline{x}^a)(\partial \overline{x}^a/\partial \overline{\overline{x}}^\alpha), \qquad (11.4a)$$

or,

$$(\partial x^i/\partial \overline{\overline{x}}^\alpha)(\partial \overline{\overline{x}}^\alpha/\partial \overline{x}^a) = (\partial x^i/\partial \overline{x}^a), \qquad (11.4b)$$

and

$$\partial \overline{\overline{x}}^\alpha/\partial x^k = (\partial \overline{\overline{x}}^\alpha/\partial \overline{x}^a)(\partial \overline{x}^a/\partial x^k). \qquad (11.4c)$$

The transformations in Eq. (11.2) further map the transformed components $\overline{X}^a$ of a contravariant vector, given by Eq. (4.1b), on to $\overline{\overline{X}}^\alpha$ as per

$$\overline{X}^a = \overline{\overline{X}}^\alpha(\partial \overline{x}^a/\partial \overline{\overline{x}}^\alpha).$$

Consequently, Eq. (4.1a) reduces to:

$$X^{i} = \overline{\overline{X}}^{\alpha} (\partial \overline{x}^{a} / \partial \overline{\overline{x}}^{\alpha})(\partial x^{i} / \partial \overline{x}^{a}) = \overline{\overline{X}}^{\alpha} (\partial x^{i} / \partial \overline{\overline{x}}^{\alpha}),$$

by Eq. (11.4a). Above equation provides a direct coordinate transformation of X^{i} onto $\overline{\overline{X}}^{\alpha}$ under the composite coordinate transformation (11.3). Thus, the composition of two coordinate transformations in Eqs. (1.2) and (11.2) is again a coordinate transformation.

Similarly, the statement can be verified for a covariant vector as well as a tensor. //

Example 11.1. The outer product of r contravariant vectors and s covariant vectors is a mixed tensor of type (r, s).

Solution. Methods employed in Examples 6.1 and 8.1 can be similarly generalized to establish the statement. //

§ 12. Some concluding remarks

Note 12.1. For simplicity, the components X^{i}, Y_{i} (forming contravariant and covariant vectors) and T^{ij}, S_{ij}, $T^{i}_{\ j}$ etc. (forming tensors) are themselves referred as respective vectors and tensors.

Note 12.2. In view of the nomenclatures in Definitions 10.1 and 11.1, the tensors in Definitions 6.1, 7.1, 8.1 and 9.1 are of types (2, 0), $(r, 0)$, (0, 2) and $(0, s)$ respectively. It is customary to write the contravariant valency as the first member and the covariant valency as the second member, while determining the type of a tensor. Also, their orders are $2, r, 2$ and s respectively.

Note 12.3. A contravariant (respectively covariant) vector is a tensor of type (1, 0) (respectively (0, 1)) while a scalar is a tensor of order zero. A vector is also called an *univalent* tensor and a scalar as *nilvalent* tensor.

Note 12.4. In addition to scalars, vectors and tensors there also exist indefinitely large number of geometric quantities obeying the coordinate transformation laws other than those mentioned above. Christoffel symbols (appearing in the next chapter) are a few of them.

§ 13. Some algebraic laws of tensors

Theorem 13.1. Tensors of same type are conformable for addition as well as subtraction; and their sum (or difference) is also a tensor of the same type.

Proof. Let A^i_j and B^i_j be two mixed tensors each of type (1, 1). So, they satisfy the coordinate transformation law vide Eq. (10.1):

$$A^i_j = \overline{A}^a_b (\partial x^i / \partial \overline{x}^a)(\partial \overline{x}^b / \partial x^j), \quad B^i_j = \overline{B}^a_b (\partial x^i / \partial \overline{x}^a)(\partial \overline{x}^b / \partial x^j).$$

Adding these relations, we get

$$A^i_j + B^i_j = (\overline{A}^a_b + \overline{B}^a_b)(\partial x^i / \partial \overline{x}^a)(\partial \overline{x}^b / \partial x^j). \tag{13.1}$$

Comparing Eq. (13.1) with Eq. (10.1), we have the statement.

For the difference of the tensors the theorem can be established similarly. //

Theorem 13.2. Multiplication of a tensor by a non-zero scalar gives a tensor of the same type.

Proof. Let $f(x^i)$ be a scalar function and T^i_j a mixed tensor of the type (1, 1). Their coordinate transformation laws are given by Eqs. (2.1) and (10.1). Multiplying these relations, we get

$$fT^i_j = (f\overline{T}^a_b)(\partial x^i / \partial \overline{x}^a)(\partial \overline{x}^b / \partial x^j);$$

which, on comparison with Eq. (10.1), expresses the coordinate transformation law of the tensor fT^i_j of type (1, 1). //

Corollary 13.1. A linear combination of the tensors of same type is again a tensor of the same type.

Proof. Let A^i_j and B^i_j be two mixed tensors of type (1, 1) and f, ϕ be two scalar functions. It follows from above theorem that fA^i_j and ϕB^i_j are also tensors of same type, i.e. (1, 1). Next, applying Theorem 13.1, we have the statement. //

Theorem 13.3. The outer product of two tensors is a tensor whose order is the sum of the orders of the given tensors.

Proof. Let $T^i_{\ j}$ and $S^k_{\ hl}$ be two mixed tensors of types (1, 1) and (1, 2) respectively. So, they satisfy the coordinate transformation law vide Eq. (10.1) and

$$S^k_{\ hl} = \bar{S}^c_{de}(\partial x^k / \partial \bar{x}^c)(\partial \bar{x}^d / \partial x^h)(\partial \bar{x}^e / \partial x^l). \qquad (13.2)$$

Multiplying Eqs. (10.1) and (13.2), we get

$$T^i_j S^k_{hl} = \bar{T}^a_b \bar{S}^c_{de}(\partial x^i/\partial \bar{x}^a)(\partial x^k/\partial \bar{x}^c)(\partial \bar{x}^b/\partial x^j)(\partial \bar{x}^d/\partial x^h)(\partial \bar{x}^e/\partial x^l),$$

which is the coordinate transformation law of a mixed tensor of type (2, 3). Thus, the outer product (tensor) has two contravariant valencies (which is the sum of contravariant valencies of the given tensors) and three covariant valencies, i.e. the sum of covariant valencies of the tensors. //

Corollary 13.2. Outer product of a contravariant tensor T^{ij} and a covariant vector Y_k is a tensor of type (2, 1).

Proof. The coordinate transformation laws of these quantities are exhibited by Eqs. (6.1) and (5.1a) respectively. Multiplying these equations, we get

$$T^{ij}\, Y_k = (\bar{T}^{ab}\, \bar{Y}_c)(\partial x^i / \partial \bar{x}^a)(\partial x^j / \partial \bar{x}^b)(\partial \bar{x}^c / \partial x^k),$$

which is the coordinate transformation law of a mixed tensor of type (2, 1). //

Note 13.1. In any tensor equation the nature of free indices remains the same on either side of the equation. As such, the equation possesses index balance.

Note 13.2. An index appearing only on one side of a tensor equation must be dummy (i.e. to be summed) on that side.

Note 13.3. A tensor with vanishing components in one coordinate system also has vanishing components in every other coordinate system.

Example 13.1. Addition of tensors of same type is commutative as well as associative.

Solution. Since the components of tensors are real numbers and the addition in the set of real numbers is commutative as well as associative, the statement also holds for tensors of same type. //

§ 14. Operations of contraction and transvection

Definition 14.1. (*Contraction*) Let $T^{i}{}_{jk}$ be a mixed tensor of type (1, 2). When the contravariant index i and a covariant index, say j, assume common values the sum of such n components of the tensor:

$$T^{i}_{ik} \equiv \sum_{i=1}^{n} T^{i}_{ik} = T^{1}_{1k} + T^{2}_{2k} + ... + T^{n}_{nk}, \tag{14.1}$$

is called the contracted part of the tensor with respect to the indices i and j. This process yielding the sum $T^{i}{}_{ik}$ is called *contraction*.

It may be noted that the indices i and j lose their free nature in above process leaving k alone as the free index. As such, the sum in Eq. (14.1) becomes a covariant vector. The following theorem describes the generalization of this rule to a tensor of arbitrary type.

Theorem 14.1. A mixed tensor of type (r, s) yields a mixed tensor of type $(r-1, s-1)$ under a process of contraction with respect to a pair of contravariant and covariant indices.

Proof. Contracting the rth contravariant index i_r and s^{th} covariant index j_s in Eq. (11.1), we get

$$T^{i_1 ... i_{r-1} i_r}_{j_1 ... j_{s-1} i_r} = \bar{T}^{a_1 ... a_{r-1} a_r}_{b_1 ... b_{s-1} b_s} (\partial x^{i_1} / \partial \bar{x}^{a_1}) ... (\partial x^{i_{r-1}} / \partial \bar{x}^{a_{r-1}}).$$

$$.(\partial x^{i_r} / \partial \bar{x}^{a_r}).(\partial \bar{x}^{b_1} / \partial x^{j_1}) ... (\partial \bar{x}^{b_{s-1}} / \partial x^{j_{s-1}}).(\partial \bar{x}^{b_s} / \partial x^{i_r})$$

$$= \bar{T}^{a_1 ... a_{r-1} a_r}_{b_1 ... b_{s-1} a_r} (\partial x^{i_1} / \partial \bar{x}^{a_1}) ... (\partial x^{i_{r-1}} / \partial \bar{x}^{a_{r-1}}).$$

$$.(\partial \bar{x}^{b_1} / \partial x^{j_1}) ... (\partial \bar{x}^{b_{s-1}} / \partial x^{j_{s-1}}),$$

where simplifications are made by means of Eq. (1.5). Above relation expresses the coordinate transformation law of a mixed tensor of type $(r - 1, s - 1)$. The dummy indices i_r and a_r on the respective sides contribute nothing towards the nature of the tensor. //

Note 14.1. The process of contraction cannot be applied to scalars or vectors.

Definition 14.2. (*Transvection*) Inner product of two tensors (having at least one pair of dummy indices - each belonging to different tensors) defines the process of *transvection.*

Thus, the transvection of a tensor T^i_{jk} by another tensor S^j_{hl} is their inner product:

$$T^i_{jk} S^j_{hl} \equiv U^i_{khl}, \tag{14.2}$$

which has a pair of dummy indices *j*'s. In the following, we establish the geometric character of the transvected part U^i_{khl} of the tensor T^i_{jk}.

Theorem 14.2. The inner product in Eq. (14.2) is a mixed tensor of type (1, 3).

Proof. Respective tensors satisfy the coordinate transformation laws

$$T^i_{jk} = \bar{T}^a_{bc} (\partial x^i / \partial \bar{x}^a)(\partial \bar{x}^b / \partial x^j)(\partial \bar{x}^c / \partial x^k), \tag{14.3}$$

and the one in Eq. (13.2). Therefore, the coordinate transformation law of their inner product is

$$T^i_{jk} S^j_{hl} = \bar{T}^a_{bc} \bar{S}^g_{de} (\partial x^i / \partial \bar{x}^a)(\partial \bar{x}^b / \partial x^j)(\partial x^j / \partial \bar{x}^g).$$

$$.(\partial \bar{x}^c / \partial x^k)(\partial \bar{x}^d / \partial x^h)(\partial \bar{x}^e / \partial x^l)$$

$$= \bar{T}^a_{bc} \bar{S}^b_{de} (\partial x^i / \partial \bar{x}^a)(\partial \bar{x}^c / \partial x^k)(\partial \bar{x}^d / \partial x^h)(\partial \bar{x}^e / \partial x^l), \tag{14.4}$$

by Eq. (1.5). As the dummy indices *j* and *b* on the respective sides of Eq. (14.4) contribute nothing towards the geometric nature, Eq. (14.4) expresses the coordinate transformation law of a mixed tensor of type (1, 3). //

On the other hand, if the tensorial character of an inner product of a geometric object with some vector (or tensor) is known the tensor character of the object can be determined by a rule - called the *quotient law* of tensors. Thus, if $T^i_{jk}(x^h)$ be a set of functions mapping onto $\bar{T}^a_{bc}(\bar{x}^d)$ under the coordinate transformations in Eq. (1.2) and its inner product with a contravariant vector with components $X^i(x^h)$ and $\bar{X}^a(\bar{x}^d)$ in respective coordinate systems be formed then the *quotient law* is stated as follows:

Theorem 14.3. If the inner product $T^i_{jk} X^k$ of some functions T^i_{jk} and a contravariant vector X^k is a tensor of order two then T^i_{jk} form a tensor of order one higher than that of the product.

Proof. As per hypothesis, the inner product $T^i_{jk} X^k \equiv A^i_j$ transforming onto $\overline{T}^a_{bc} \overline{X}^c \equiv \overline{A}^a_b$ is a mixed tensor of order two and X^k is a contravariant vector, there holds the transformation law

$$T^i_{jk} X^k = (\overline{T}^a_{bc} \overline{X}^c)(\partial x^i / \partial \overline{x}^a)(\partial \overline{x}^b / \partial x^j),$$

or, for Eq. (4.1 a),

$$\{T^i_{jk} (\partial x^k / \partial \overline{x}^c) - \overline{T}^a_{bc} (\partial x^i / \partial \overline{x}^a) (\partial \overline{x}^b / \partial x^j)\}\overline{X}^c = 0.$$

$\overline{X}^c$ being arbitrary, above equation implies

$$T^i_{jk} (\partial x^k / \partial \overline{x}^c) = \overline{T}^a_{bc} (\partial x^i / \partial \overline{x}^a) (\partial \overline{x}^b / \partial x^j).$$

On transvection by $\partial \overline{x}^c / \partial x^h$, for Eq. (1.4), above equation determines the coordinate transformation law as per Eq. (14.3) proving that T^i_{jk} is a mixed tensor of type (1, 2). //

Note 14.2. The transvection process applies between two tensors having at least one pair of dummy indices - each belonging to different tensors; while the process of contraction is applied to a single mixed tensor by making a pair of one of its contravariant and the other covariant indices dummy.

Note 14.3. Together with Note 14.1, we also note that these processes can also be applied to geometric objects other than tensors.

Theorem 14.4. The inner (or scalar) product of a contravariant vector X^i and a covariant vector Y_i is a scalar.

Proof. The respective vectors obey the coordinate transformation laws exhibited by Eqs. (4.1a) and (5.1a). Accordingly, their inner product transforms as

$$X^i Y_i = (\bar{X}^a \bar{Y}_b)(\partial x^i / \partial \bar{x}^a)(\partial \bar{x}^b / \partial x^i) = \bar{X}^a \bar{Y}_a, \tag{14.5}$$

by Eq. (1.5). Comparing Eq. (14.5) with Eq. (2.1) we conclude the invariant character of the product. //

Definition 14.3. A contravariant tensor S^{ij} and a covariant tensor S_{jk} satisfying

$$S^{ij} S_{jk} = \delta^i_k \tag{14.6}$$

are called the *reciprocal tensors*.

Theorem 14.5. A tensor equation remains invariant under a coordinate transformation.

Proof. Let

$$T^i_j + A^i B_j - \psi S^i_{jk} X^k = 0 \tag{14.7}$$

be a tensor equation, where T^i_j and S^i_{jk} are mixed tensors of types (1, 1) and (1, 2) respectively; A^i, X^k contravariant vectors; B_j a covariant vector and ψ is a scalar. In view of Example 10.1 and Theo. 13.2 and 14.2, the latter two terms in the equation are also tensors of type (1, 1). Hence, by Theo. 13.1, the sum, in Eq. (14.7) is also a tensor of the same type. As this tensor is zero in the original coordinate system (x^i), for Note 13.3, it remains zero in the transformed coordinate system ($\bar{x}^a$) as well. Thus, applying the coordinate transformations of respective quantities, the Eq. (14.7) transforms as

$$\{\bar{T}^a_b + \bar{A}^a \bar{B}_b - \psi \bar{S}^a_{bc}(\partial \bar{x}^c / \partial x^k)\bar{X}^d (\partial x^k / \partial \bar{x}^d)\}.$$

$$.(\partial x^i / \partial \bar{x}^a)(\partial \bar{x}^b / \partial x^j) = 0,$$

or, for Eq. (1.5),

$$\{\bar{T}^a_b + \bar{A}^a \bar{B}_b - \psi \bar{S}^a_{bc} \bar{X}^c\}(\partial x^i / \partial \bar{x}^a)(\partial \bar{x}^b / \partial x^j) = 0.$$

Its transvection by $(\partial \bar{x}^d / \partial x^i)$ $(\partial x^j / \partial \bar{x}^e)$ and applications of Eq. (1.5) yields

$$\bar{T}^d_e + \bar{A}^d \bar{B}_e - \psi \bar{S}^d_{ec} \bar{X}^c = 0,$$

which is same as Eq. (14.7). //

Example 14.1. An inner product of the tensor T^{ij} and S_l^{kh} is a contravariant tensor of order three.

Solution. The tensors satisfy the coordinate transformation laws vide Eq. (6.1) and

$$S_l^{kh} = \overline{S}_e^{cd}(\partial x^k / \partial \overline{x}^c)(\partial x^h / \partial \overline{x}^d)(\partial \overline{x}^e / \partial x^l). \quad (14.8)$$

Inner products of the tensors can be computed by contracting a pair of indices of opposite valencies in the respective tensors. Since T^{ij} does not possess any covariant valency, any of its contravariant valency, say i, be contracted with the only covariant valency l in the second tensor. Thus, for Eqs. (6.1) and (14.8), the inner product $T^{ij} S_i^{kh}$ satisfies

$$T^{ij} S_i^{kh} = (\overline{T}^{ab}\,\overline{S}_e^{cd})(\partial x^i / \partial \overline{x}^a)(\partial \overline{x}^e / \partial x^i)(\partial x^j / \partial \overline{x}^b)(\partial x^k / \partial \overline{x}^c).$$
$$.(\partial x^h / \partial \overline{x}^d) = (\overline{T}^{ab}\,\overline{S}_a^{cd})(\partial x^j / \partial \overline{x}^b)(\partial x^k / \partial \overline{x}^c)(\partial x^h / \partial \overline{x}^d), \quad (14.9)$$

where simplifications are made by means of Eq. (1.5). We note that Eq. (14.9) expresses the coordinate transformation law of a contravariant tensor of order three. As seen in Theo. 14.1, the contracted index i in the inner product contributes nothing towards the geometric nature of the tensor $T^{ij} S_i^{kh}$. //

Example 14.2. Inner products of a contravariant tensor T^{ij} and a covariant vector Y_k are contravariant vectors.

Solution. As explained in the previous example, one of the inner product is $T^{ij} Y_i$; which, for Eqs. (5.1a) and (6.1) satisfies.

$$T^{ij} Y_i = (\overline{T}^{ab}\,\overline{Y}_c)(\partial x^i / \partial \overline{x}^a)(\partial x^j / \partial \overline{x}^b)(\partial \overline{x}^c / \partial x^i)$$
$$= (\overline{T}^{ab}\,\overline{Y}_a)(\partial x^j / \partial \overline{x}^b), \quad (14.10)$$

by Eq. (1.5). Putting

$$T^{ij} Y_i \equiv A^j, \quad \text{and} \quad \overline{T}^{ab}\,\overline{Y}_a = \overline{A}^b;$$

and comparing Eq. (14.10) with Eq. (5.1a), we have the statement. //

Example 14.3. Inner products of a covariant tensor S_{ij} and a contravariant vector X^i are covariant vectors.

Solution. For Eqs. (4.1a) and (8.1), the inner product $S_{ij}\,X^i$ satisfies the coordinate transformation law

$$S_{ij}X^i = \overline{S}_{ab}\,\overline{X}^c\,(\partial\overline{x}^a/\partial x^i)(\partial\overline{x}^b/\partial x^j)(\partial x^i/\partial\overline{x}^c)$$

$$= \overline{S}_{ab}\,\overline{X}^a\,(\partial\overline{x}^b/\partial x^j), \qquad (14.11)$$

by Eq. (1.5). A comparison of Eq. (14.11) with Eq. (5.1a) establishes the statement. //

§ 15. Symmetric and skew-symmetric tensors

The components S_{ij} of a tensor are generally independent to each other. However, if they satisfy

$$S_{ij} = S_{ji} \qquad (15.1);\ \text{(respectively)} \qquad S_{ij} = -S_{ji}, \qquad (15.2)$$

$\forall\ i, j = 1, 2, \ldots, n$ the tensor is called *symmetric* (respectively *skew-symmetric*). This concept can also be generalized to a tensor of arbitrary order.

Definition 15.1. The tensor defined by Eq. (9.1) is called *symmetric* (respectively *skew-symmetric*) with respect to its pth and q^{th} indices if it equals positive (resp. negative) of

$$S_{j_1 \cdots j_{p-1}\, j_q\, j_{p+1} \cdots j_{q-1}\, j_p\, j_{q+1} \cdots j_s}, \qquad \forall\ j_p, j_q = 1, 2, \ldots, n.$$

Theorem 15.1. A symmetric tensor of second order has at most $n\,(n+1)/2$ linearly independent components in a space V_n.

Proof. Let a_{ij} be the components of a symmetric covariant tensor of order two. Writing all its n^2 components:

$$\left.\begin{array}{cccccc}
a_{11}, & a_{12}, & \ldots, & a_{1j}, & \ldots, & a_{1n}; \\
a_{21}, & a_{22}, & \ldots, & a_{2j}, & \ldots, & a_{2n}; \\
\cdot\,, & \cdot\,, & \ldots, & \cdot\,, & \ldots, & \cdot\,; \\
a_{i1}, & a_{i2}, & \ldots, & a_{ij}, & \ldots, & a_{in}; \\
\cdot\,, & \cdot\,, & \ldots, & \cdot\,, & \ldots, & \cdot\,; \\
a_{n1}, & a_{n2}, & \ldots, & a_{nj}, & \ldots, & a_{nn};
\end{array}\right\} \qquad (15.3)$$

we note that the maximum number of row-wise linearly independent components are:

Row	No. of components	Reason
1	n	
2	$n-1$	$a_{21} = a_{12}$ is already counted in first row.
3	$n-2$	$a_{31} = a_{13}, a_{32} = a_{23}$ are counted earlier.
…	…	….
n	1	Only a_{nn}, rest ones are already counted earlier.

Thus, the total number of such components is

$$n + (n-1) + (n-2) + \ldots + 1 \;=\; \sum n = n\,(n+1)\,/\,2. \;//$$

Theorem 15.2. Number of linearly independent components of a skew-symmetric tensor of second order cannot exceed $n\,(n-1)/2$ in V_n.

Proof. Let a_{ij} form the components of a skew-symmetric tensor of order two. In view of Eq. (15.2), its components satisfy

$$a_{ij} \;=\; 0 \;\;(\text{when } i = j), \quad -a_{ji} \;(\text{when } i \neq j). \tag{15.4}$$

Accordingly, its components are

$$\begin{array}{cccccc}
0\,, & a_{12}, & \ldots, & a_{1j}, & \ldots, & a_{1n}; \\
a_{21}, & 0\,, & \ldots, & a_{2j}, & \ldots, & a_{2n}; \\
.\,, & .\,, & \ldots, & .\,, & \ldots, & .\;; \\
a_{i1}, & a_{i2}, & \ldots, & a_{ij}, & \ldots, & a_{in}; \\
.\,, & .\,, & \ldots, & .\,, & \ldots, & .\;; \\
a_{n1}, & a_{n2}, & \ldots, & a_{nj}, & \ldots, & 0\;.
\end{array}$$

Hence, the maximum number of row-wise linearly independent components are

$$(n-1)+(n-2)+\dots+1+0=\sum(n-1)=n(n-1)/2. \;//$$

Theorem 15.3. Any geometric entity S_{ij} can be decomposed as the sum of a symmetric and a skew-symmetric parts.

Proof. Defining

$$S_{(ij)} \equiv (1/2)(S_{ij}+S_{ji}), \tag{15.5}$$

and

$$S_{[ij]} \equiv (1/2)(S_{ij}-S_{ji}), \tag{15.6}$$

and interchanging the indices i and j, above relations establish

$$S_{(ij)} = S_{(ji)}, \qquad \text{and} \qquad S_{[ij]} = -S_{[ji]},$$

$\forall\, i, j = 1, 2, \dots, n$. Thus, for Eqs. (15.1) and (15.2), the quantities in (15.5) and (15.6) are symmetric and skew-symmetric respectively. Further, addition of Eqs. (15.5) and (15.6) yields

$$S_{ij} = S_{(ij)} + S_{[ij]}, \tag{15.7}$$

that proves the statement. //

Note 15.1. The geometric entities in Eqs. (15.5) and (15.6) are called the *symmetric* and *skew-symmetric parts of* S_{ij}.

Example 15.1. For any tensor a_{ij} the tensors

$$a_{(ij)} \equiv (1/2)(a_{ij}+a_{ji}) \quad (15.8);\ \text{and} \quad a_{[ij]} \equiv (1/2)(a_{ij}-a_{ji}) \quad (15.9)$$

are respectively symmetric and skew-symmetric.

Solution. Proof is similar to that of Theo. 15.3. //

Example 15.2. The tensor

$$S_{ij} \equiv A_i B_j + A_j B_i, \tag{15.10}$$

where A_i, B_j are covariant vectors, is symmetric and is of rank two.

Solution. Interchanging the indices i and j in Eq. (15.10), we get

$$S_{ji} = A_j B_i + A_i B_j = S_{ij};$$

which establishes the symmetry of the tensor.

Further, the components of the tensor form the matrix

$$[S_{ij}] = \begin{bmatrix} A_1 B_1 + A_1 B_1 & A_1 B_2 + A_2 B_1 & \dots & A_1 B_n + A_n B_1 \\ A_2 B_1 + A_1 B_2 & A_2 B_2 + A_2 B_2 & \dots & A_2 B_n + A_n B_2 \\ \cdot & \cdot & \dots & \cdot \\ A_n B_1 + A_1 B_n & A_n B_2 + A_2 B_n & \dots & A_n B_n + A_n B_n \end{bmatrix}$$

$$= \begin{bmatrix} A_1 B_1 & A_1 B_2 & \dots & A_1 B_n \\ A_2 B_1 & A_2 B_2 & \dots & A_2 B_n \\ \cdot & \cdot & \dots & \cdot \\ A_n B_1 & A_n B_2 & \dots & A_n B_n \end{bmatrix} + \begin{bmatrix} A_1 B_1 & A_2 B_1 & \dots & A_n B_1 \\ A_1 B_2 & A_2 B_2 & \dots & A_n B_2 \\ \cdot & \cdot & \dots & \cdot \\ A_1 B_n & A_2 B_n & \dots & A_n B_n \end{bmatrix}.$$

Taking out common multiples in respective columns either of above matrices can be seen singular. As in Example 6.2, every minor of any order from 2 to $n - 1$, in the determinants of above matrices vanishes identically. Thus, either of above matrices are of rank one. Hence,

$$\text{rank}\,((S_{ij})) = 1 + 1 = 2. \ //$$

Example 15.3. For scalars a, b and a tensor S_{ij} the tensor equation

$$a\, S_{ij} + b\, S_{ji} = 0, \tag{15.11}$$

implies

(i) either $a = -b$ and symmetry of the tensor ;

(ii) or $a = b$ and skew-symmetry of the tensor.

Solution. Interchanging the indices i, j in Eq. (15.11) and adding the resulting equation to Eq. (15.11), there follows

$$(a + b)\ (S_{ij} + S_{ji}) = 0\ ;$$

which necessarily concludes

(i) either $a + b = 0$ so that $a = -b$ together with Eq. (15.11) causing symmetry of S_{ij} ;

(ii) or, $S_{ij} + S_{ji} = 0 \quad \Rightarrow$ skew-symmetry of the tensor.

Accordingly, Eq. (15.11) implies $a - b = 0$, or $a = b$. //

§ 16. Relative tensors

Referring back to Section 11, if the transformation law of the functions $T^{i_1 i_2 \dots i_r}_{j_1 j_2 \dots j_s}$ includes an additional factor $\left| \partial \bar{x}^a / \partial x^i \right|^w$ in the right side of Eq. (11.1), the functions $T^{\cdots}_{\cdots}$ are then said to form a *relative tensor* of weight w, where the Jacobian

$$\frac{\partial\,(\bar{x}^1,\bar{x}^2,\dots,\bar{x}^n)}{\partial(x^1,x^2,\dots,x^n)} \equiv \det\left(\frac{\partial\bar{x}^a}{\partial x^i}\right) = \begin{vmatrix} \dfrac{\partial\bar{x}^1}{\partial x^1} & \dfrac{\partial\bar{x}^1}{\partial x^2} & \cdots & \dfrac{\partial\bar{x}^1}{\partial x^n} \\ \dfrac{\partial\bar{x}^2}{\partial x^1} & \dfrac{\partial\bar{x}^2}{\partial x^2} & \cdots & \dfrac{\partial\bar{x}^2}{\partial x^n} \\ \cdot & \cdot\,\cdot & \cdots & \cdot \\ \dfrac{\partial\bar{x}^n}{\partial x^1} & \dfrac{\partial\bar{x}^n}{\partial x^2} & \cdots & \dfrac{\partial\bar{x}^n}{\partial x^n} \end{vmatrix} \neq 0. \tag{16.1}$$

Definition 16.1. A relative tensor of weight zero (respectively one) is called on *absolute tensor* (respectively *tensor density*).

Note 16.1. All the geometric quantities defined in the foregoing sections are, thus, absolute ones.

Unless stated otherwise, all the geometric quantities appearing in our discussion will be of weight zero.

Example 16.1. The quantities ε^{ijk} defined by

$$\varepsilon^{ijk} \equiv \begin{cases} 1 & \text{if } i, j, k \text{ form an even permutation of } 1,2,3\,; \\ -1 & \text{if } i, j, k \text{ form an odd permutation of } 1,2,3\,; \\ 0 & \text{otherwise, i.e. when at least two of } i, j, k \\ & \text{take equal values}\,; \end{cases} \tag{16.2}$$

form the components of a contravariant tensor of order 3 and weight one. This tensor is called the *permutation tensor*.

Solution. Putting $p^i_a \equiv \partial x^i / \partial \bar{x}^a$ for brevity, the determinant

$$\left|\frac{\partial x^i}{\partial \bar{x}^a}\right| = \begin{vmatrix} \partial x^1/\partial \bar{x}^1 & \partial x^1/\partial \bar{x}^2 & \partial x^1/\partial \bar{x}^3 \\ \partial x^2/\partial \bar{x}^1 & \partial x^2/\partial \bar{x}^2 & \partial x^2/\partial \bar{x}^3 \\ \partial x^3/\partial \bar{x}^1 & \partial x^3/\partial \bar{x}^2 & \partial x^3/\partial \bar{x}^3 \end{vmatrix}$$

$$= p_1^1\,(p_2^2 p_3^3 - p_3^2 p_2^3) + p_2^1(p_3^2 p_1^3 - p_1^2 p_3^3) + p_3^1\,(p_1^2 p_2^3 - p_2^2 p_1^3)$$

$$= \sum \pm\, p_a^1\; p_b^2\; p_c^3\,,$$

where positive (respectively negative) sign is accounted when the indices a, b, c form an even (respectively odd) permutation of $1, 2, 3$. Defining the transformed counterparts $\bar{\varepsilon}^{abc}$ of ε^{ijk} under the coordinate transformation vide Eq. (1.2) in analogy with Eq. (16.2), above determinant can also be written as:

$$\left|\partial x^i / \partial \bar{x}^a\right| = \bar{\varepsilon}^{abc}\; p_a^1\; p_b^2\; p_c^3\,, \tag{16.3}$$

where a, b, c are summed over $1, 2, 3$. Above equation also implies

$$\bar{\varepsilon}^{abc}\; p_a^i\; p_b^j\; p_c^k \;=\; \pm\left|\partial x^i / \partial \bar{x}^a\right| \;=\; \varepsilon^{ijk}\;\left|\partial x^i / \partial \bar{x}^a\right|,$$

where positive (respectively negative) sign is taken when i, j, k form an even (respectively odd) permutation of $1, 2, 3$. Rewriting above equation as

$$\varepsilon^{ijk} = \left|\partial x^i / \partial \bar{x}^a\right|^{-1} \bar{\varepsilon}^{abc}\; p_a^i\; p_b^j\; p_c^k = \left|\partial \bar{x}^a / \partial x^i\right| \bar{\varepsilon}^{abc}\; p_a^i\; p_b^j\; p_c^k \tag{16.4}$$

which proves the statement. //

§ 17. Problem set

17.1. Let T^{ij} be a contravariant tensor of second order. Show that the functions

$$B^{ij} \equiv T^{ji} \tag{17.1}$$

also form a tensor of the same type.

[**Hint.** Interchanging the indices i, j in Eq. (6.1) and putting from Eq. (17.1), there follows

$$B^{ij} = \overline{T}^{ab}\,(\partial x^j/\partial \overline{x}^a)(\partial x^i/\partial \overline{x}^b) = \overline{T}^{ba}\,(\partial x^j/\partial \overline{x}^b)(\partial x^i/\partial \overline{x}^a)$$

$$= \overline{B}^{ab}\,(\partial x^i/\partial \overline{x}^a)(\partial x^j/\partial \overline{x}^b),$$

which proves the statement.]

17.2. Let S_{ij} and X^j form a tensor and a vector respectively satisfying

$$|S_{ij}| = 0, \quad \text{and} \quad S_{ij}X^j = 0. \tag{17.2}$$

Show that there exists a coordinate system $(\overline{x}^a)$ for which the transformed components of the tensor vanish identically.

[**Hint.** Using Eqs. (4.1a) and (8.1), the second relation in Eqs. (17.2) transforms under the coordinate transformation vide Eq. (1.2) as

$$\left.\begin{aligned} &\overline{S}_{ab}(\partial \overline{x}^a/\partial x^i)(\partial \overline{x}^b/\partial x^j)\overline{X}^c(\partial x^j/\partial \overline{x}^c) \\ &= \overline{S}_{ab}\,\overline{X}^b(\partial \overline{x}^a/\partial x^i) = 0, \end{aligned}\right\} \tag{17.3}$$

where Eq. (1.5) is also used. The statement follows from Eq. (17.3) as $\overline{X}^b$ and $\partial \overline{x}^a/\partial x^i$ are non-zero.]

17.3. Show that the extended Kronecker deltas

$$\delta^{ij}_{kh} \equiv \begin{cases} 1 & \text{if} \quad i = k \neq h = j, \\ -1 & \text{if} \quad i = h \neq k = j, \\ 0 & \text{otherwise}; \end{cases} \tag{17.4}$$

form the components of an absolute tensor of order four.

17.4. Show that the extended Kronecker deltas given by equation (17.4) satisfy

$$\delta^{ij}_{kh} = \delta^i_k\,\delta^j_h - \delta^i_h\,\delta^j_k. \tag{17.5}$$

[**Hint.** Denoting the right member of Eq. (17.5) by A^{ij}_{kh}, we note that:

(i) when $k = h$, the tensor A^{ij}_{kh} is clearly zero ;

(ii) when $k \neq h$, say $k = i$ so that $h \neq i \Rightarrow \delta^i{}_h = 0,\ \delta^i{}_k = 1$.

Accordingly, $A^{ij}_{kh} = \delta^j{}_h = 1$ when $h = j$, otherwise zero.

(iii) when $k \neq h$ and $k \neq i \Rightarrow \delta^i{}_k = 0$ so that

$$A^{ij}_{kh} = -\,\delta^i{}_h\,\delta^j{}_k = \begin{cases} -1 & \text{when} \quad h = i,\ \ j = k; \\ \ \ 0 & \text{when} \quad h \neq i \ \text{ and/or } \ k \neq j. \end{cases}$$

Thus, the values of A^{ij}_{kh} coincide with those of δ^{ij}_{kh} as defined by Eq. (17.4).]

17.5. Show that

$$\delta^{ij}_{kh}\ S_{ij} \ = \ S_{kh} - S_{hk}\ , \qquad (17.6)$$

where S_{ij} is any geometric object.

[**Hint.** Use the identity (17.5).]

17.6. Prove that the inner product of two tensors a_{ij} and b^{kh} is a tensor of type (1,1).

[**Hint.** Contraction with respect to different pairs of indices yield the inner products

$$A^h_j = a_{ij}\ b^{ih}, \quad B^k_j \equiv a_{ij}\ b^{ki}, \quad C^k_i \equiv a_{ij}\ b^{kj}, \quad D^h_i \equiv a_{ij}\ b^{jh}\,.$$

Show that each of these products yield tensors of type (1, 1).]

17.7. If $A^i\,B^j\,C_{ij}$ is a scalar invariant and A^i, B^j are contravariant vectors then show that C_{ij} form a second order covariant tensor.

17.8. If S^{ij} and S_{ij} are reciprocal symmetric tensors and

$$Y_i \equiv S_{ij}\ Y^j \qquad (17.7)$$

is a covariant vector than show that

$$S^{ij}\,Y_i\,Y_j = S_{ij}\ Y^i\,Y^j. \qquad (17.8)$$

[**Hint.** Putting from Eq. (17.7), the left member of Eq. (17.8) becomes

$$S^{ij}(S_{ik}Y^k)(S_{jh}Y^h) = \delta^j_k Y^k S_{jh} Y^h = Y^j S_{jh} Y^h,$$

by Eq. (14.6).]

17.9. Reduce the tensor equation

$$T^i_{\ j} Y_i = \alpha Y_j \tag{17.9}$$

to the form

$$(T^i_{\ j} - \alpha\, \delta^i_j) Y_i = 0, \tag{17.10}$$

where α is an invariant. Further, show that

$$T^i_{\ j} = \alpha\, \delta^i_j, \tag{17.11}$$

if Eq. (17.10) holds for an arbitrary vector Y_i.

[**Hint.** For $Y_j = \delta^i_j Y_i$, Eq. (17.9) assumes the form of Eq. (17.10). Further, Eq. (17.11) results from Eq. (17.10) for arbitrary Y_i's.]

17.10. If there holds the equation (17.9) for all vectors Y_i satisfying

$$X^i Y_i = 0, \tag{17.12}$$

where X^i is a given vector, then show that

$$T^i_{\ j} = \alpha\, \delta^i_j + \sigma_j X^i, \tag{17.13}$$

where σ_j is some covariant vector.

[**Hint.** As in the previous problem, Eq. (17.9) reduces to Eq. (17.10). Hence, subtraction of σ_j multiple of Eq. (17.12) from Eq. (17.10), yields

$$(T^i_{\ j} - \alpha\, \delta^i_j - \sigma_j X^i) Y_i = 0,$$

which implies Eq. (17.13) for arbitrary Y_i.]

17.11. Show that the tensor equation

$$a_{ij} x^i x^j = b_{ij} x^i x^j \tag{17.14}$$

has solution

$$a_{ij} + a_{ji} = b_{ij} + b_{ji} \tag{17.15}$$

for arbitrary x^i's. Further, if the tensors a_{ij} and b_{ij} are symmetric then they are equal.

[**Hint.** Putting $A_{ij} \equiv a_{ij} - b_{ij}$, the Eq. (17.14) reduces to Eq. (8.3) having solution vide Eq. (8.4) :

$$A_{ij} + A_{ji} = a_{ij} - b_{ij} + a_{ji} - b_{ji} = 0,$$

which is the same as Eq. (17.15). Next, symmetry of the tensors reduces Eq. (17.15) to $a_{ij} = b_{ij}$.]

17.12. Show that the equations

$$(a_{ij} - k\, g_{ij})\lambda^i = 0 = (a_{ij} - k'\, g_{ij})\mu^i, \qquad k \neq k', \tag{17.16}$$

satisfied by two symmetric tensors a_{ij} and g_{ij}, and contravariant vectors λ^i and μ^i imply

$$g_{ij}\, \lambda^i \mu^j = a_{ij}\, \lambda^i \mu^j = 0, \tag{17.17}$$

and

$$k = a_{ij}\, \lambda^i \lambda^j / g_{lm}\, \lambda^l \lambda^m. \tag{17.18}$$

[**Hint.** Multiplying the respective equations in Eq. (17.16) by μ^j and λ^j, taking the difference of the resulting equations and using symmetry of given tensors, there follow Eqs. (17.17). Further, a transvection of the first of Eqs. (17.16) by λ^j determines Eq. (17.18).]

CHAPTER 5

TENSORS IN CYLINDRICAL AND SPHERICAL COORDINATES

§ 1. Orthogonal curvilinear coordinates

We consider a three-dimensional Euclidean space E_3 equipped with some origin O and the rectangular Cartesian coordinate axes Ox^1, Ox^2, Ox^3. Let the unit vectors $\hat{\mathbf{e}}_i$ [3] acting along the respective coordinate axes form a basis of the space. Let P be a point in the space with rectangular Cartesian Coordinates $(x^i) \equiv (x^1, x^2, x^3)$ so that its position vector $\overrightarrow{OP} \equiv \mathbf{r}$ with respect to above basis is given by

$$\mathbf{r} = x^i\, \hat{\mathbf{e}}_i \equiv (x^1, x^2, x^3), \qquad (1.1)$$

where the Einstein's summation convention is used.

Extending the concept of V_2 (cf. [12], Chap. 15) we review the space E_3 with regard to a curvilinear coordinate system having three independent parameters. When the coordinates x^i's are expressible as single-valued differentiable functions of three parameters u^α's:

$$x^i = x^i(u^\alpha) \equiv x^i(u^1, u^2, u^3), \qquad (1.2)$$

together with unique solution of above equations :

$$u^\alpha = u^\alpha(x^i) = u^\alpha(x^1, x^2, x^3); \qquad (1.3)$$

the parameters u^α's form a curvilinear coordinate system in E_3. There exist the level surfaces

$$u^\alpha = u^\alpha_0, \qquad (1.4)$$

called the *coordinate surfaces* through an arbitrary point $P_0(u^\alpha_0)$. Pairs of these surfaces intersect each other in curves called the *parametric curves* or (curvilinear) coordinate curves. For instance, the surfaces

---

3) Throughout this chapter both the Greek and Latin indices take values 1,2,3 only. Rectangular Cartesian coordinates are assigned Latin indices while the curvilinear coordinates Greek indices.

$u^1 = u^1_0$, and $u^2 = u^2_0$, intersect in the u^3-curve, as the only variable parameter along this curve is u^3. Similarly, the other pairs of surfaces:

$$u^2 = u^2_0, \quad u^3 = u^3_0, \quad \text{and} \quad u^3 = u^3_0, \quad u^1 = u^1_0,$$

determine the u^1- and u^2-curves respectively. If the coordinate surfaces intersect each other orthogonally the parametric curves are orthogonal and the parameters u^α's are called *orthogonal* (*curvilinear*) *coordinates*. Henceforth, our discussion employs the orthogonal curvilinear coordinates only.

Let Q ($\mathbf{r} + d\mathbf{r}$) be a neighbouring point of P ($\mathbf{r}$) so that the infinitesimal vector

$$\overrightarrow{PQ} = d\mathbf{r} = (dx^i)\,\hat{\mathbf{e}}_i = (dx^1, dx^2, dx^3), \tag{1.5}$$

or, by Taylor's theorem,

$$d\mathbf{r} = \mathbf{r}_{,\alpha}\, du^\alpha, \tag{1.6}$$

where the comma followed by an index denotes the partial derivative with respect to the variable: curvilinear coordinate if Greek index is used and rectangular Cartesian coordinate if a Latin index is used. The vectors $\mathbf{r}_{,\alpha}$ are tangential to the curvilinear coordinate curves at the point P. Forming the scalar product of Eq. (1.5) with the gradient vector

$$\nabla u^\alpha \equiv u^\alpha_{,i}\, \hat{\mathbf{e}}_i = (\partial u^\alpha / \partial x^1, \partial u^\alpha / \partial x^2, \partial u^\alpha / \partial x^3),^{4)} \tag{1.7}$$

we derive

$$\nabla u^\alpha . d\mathbf{r} \equiv (u^\alpha_{,i}\, \hat{\mathbf{e}}_i) . (\hat{\mathbf{e}}_j\, dx^j) = u^\alpha_{,i}\, dx^i = du^\alpha, \tag{1.8}$$

where the identities

$$\hat{\mathbf{e}}_i . \hat{\mathbf{e}}_j = \delta_{ij} \tag{1.9}$$

satisfied by the orthonormal vectors $\hat{\mathbf{e}}_i$'s are used in association with [12], Eq. (15.2.18). On the other hand, the scalar product of ∇u^α with Eq. (1.6) is

$$\nabla u^\alpha . d\mathbf{r} = (\nabla u^\alpha . \mathbf{r}_{,\beta})\, du^\beta. \tag{1.10}$$

Hence, from Eqs. (1.8) and (1.10), follows

---

4) Here, the index balance is not preserved.

$$du^{\alpha} \;=\; (\nabla u^{\alpha}\,.\,\mathbf{r},_{\beta})\, du^{\beta}, \tag{1.11}$$

determining

$$\left.\begin{array}{lll} \nabla u^{1}\,.\,\mathbf{r},_{1} = 1, & \nabla u^{1}\,.\,\mathbf{r},_{2} = 0, & \nabla u^{1}\,.\,\mathbf{r},_{3} = 0, \\ \nabla u^{2}\,.\,\mathbf{r},_{1} = 0, & \nabla u^{2}\,.\,\mathbf{r},_{2} = 1, & \nabla u^{3}\,.\,\mathbf{r},_{3} = 0, \\ \nabla u^{3}\,.\,\mathbf{r},_{1} = 0, & \nabla u^{3}\,.\,\mathbf{r},_{2} = 0, & \nabla u^{3}\,.\,\mathbf{r},_{3} = 1, \end{array}\right\} \tag{1.12a}$$

or, equivalently,

$$\nabla u^{\alpha}\,.\,\mathbf{r},_{\beta} \;=\; \delta^{\alpha}_{\beta}\,. \tag{1.12b}$$

Hence, the sets of vectors ∇u^{α} and $\mathbf{r},_{\alpha}$ constitute reciprocal systems of vectors determining

$$\nabla u^{\alpha} \;=\; (\,\mathbf{r},_{\beta} \times \mathbf{r},_{\gamma})\,/\,K, \tag{1.13}$$

where α, β, γ form a cycle of 1, 2, 3; and

$$K \;=\; [\mathbf{r},_{1}\;\; \mathbf{r},_{2}\;\; \mathbf{r},_{3}] \tag{1.14}$$

is the scalar triple product of the vectors $\mathbf{r},_{\alpha}$'s.

Let the unit tangent vectors to the respective orthogonal curvilinear coordinate curves be denoted by $\hat{\mathbf{t}}^{\alpha}$:

$$\hat{\mathbf{t}}^{\alpha} \;=\; \mathbf{r},_{\alpha}\,/\,|\,\mathbf{r},_{\alpha}| \;\equiv\; \mathbf{r},_{\alpha}\,/\,h_{\alpha}. \tag{1.15}$$

Since the orthonormal vectors $\hat{\mathbf{t}}^{\alpha}$'s satisfy

$$[\hat{\mathbf{t}}^{1}\;\; \hat{\mathbf{t}}^{2}\;\; \hat{\mathbf{t}}^{3}] \;=\; 1, \tag{1.16}$$

there holds

$$K = h_{1}\, h_{2}\, h_{3}. \tag{1.17}$$

Accordingly, Eq. (1.13) reduces to

$$\nabla u^{\alpha} \;=\; (h_{\beta}\, h_{\gamma}\;/\,h_{1}\, h_{2}\, h_{3})(\hat{\mathbf{t}}^{\beta} \times \hat{\mathbf{t}}^{\gamma}), \tag{1.18}$$

and determines

$$\left.\begin{array}{l} \nabla u^{1} \;=\; (\hat{\mathbf{t}}^{2} \times \hat{\mathbf{t}}^{3})/h_{1} = \hat{\mathbf{t}}^{1}/h_{1}, \quad \nabla u^{2} \;=\; (\hat{\mathbf{t}}^{3} \times \hat{\mathbf{t}}^{1})/h_{2} = \hat{\mathbf{t}}^{2}/h_{2}, \\ \nabla u^{3} \;=\; (\hat{\mathbf{t}}^{1} \times \hat{\mathbf{t}}^{2})/h_{3} = \hat{\mathbf{t}}^{3}/h_{3}; \end{array}\right\} \tag{1.19}$$

so that

$$[\nabla u^1 \ \nabla u^2 \ \nabla u^3] = [\hat{\mathbf{t}}^1 \ \hat{\mathbf{t}}^2 \ \hat{\mathbf{t}}^3]/K = 1/K, \qquad (1.20)$$

by Eqs. (1.16) and (1.17).

§ 2. ∇ operator in orthogonal curvilinear coordinates

Let $f(x^i)$ be a scalar point function of rectangular Cartesian coordinates which remains invariant under the coordinate transformation in Eq. (1.2):

$$f(x^i) = f(u^\alpha). \qquad (2.1)$$

For Eq. (1.3), the partial derivatives of f with respect to x^i's are given by

$$f_{,i} = f_{,\alpha}\, u^\alpha{}_{,i}\,. \qquad (2.2)$$

Multiplying Eq. (2.2) by $\hat{\mathbf{e}}_i$'s and performing summation over the index i, we derive

$$\nabla f = f_{,i}\,\hat{\mathbf{e}}_i = f_{,\alpha}\,(u^\alpha_{,i}\ \hat{\mathbf{e}}_i) = f_{,\alpha}\,\nabla u^\alpha \qquad (2.3a)$$

$$= (f_{,1}/h_1)\,\hat{\mathbf{t}}^1 + (f_{,2}/h_2)\,\hat{\mathbf{t}}^2 + (f_{,3}/h_3)\,\hat{\mathbf{t}}^3; \qquad (2.3b)$$

by Eqs. (1.7) and (1.19). Above result describes the gradient of f in the orthogonal curvilinear coordinates u^α's.

Theorem 2.1. The divergence of a vector point function

$$\mathbf{F}(u^\alpha) \equiv f_\beta\,\hat{\mathbf{t}}^\beta \qquad (2.4)$$

in the orthogonal curvilinear coordinates (u^α) is given by

$$\nabla \cdot \mathbf{F} = \{(f_1 h_2 h_3)_{,1} + (f_2 h_3 h_1)_{,2} + (f_3 h_1 h_2)_{,3}\} / K. \qquad (2.5)$$

Proof. Rewriting Eq. (1.19) as

$$\left.\begin{aligned} \hat{\mathbf{t}}^1 &= \hat{\mathbf{t}}^2 \times \hat{\mathbf{t}}^3 = h_2 h_3 (\nabla u^2 \times \nabla u^3), \\ \hat{\mathbf{t}}^2 &= \hat{\mathbf{t}}^3 \times \hat{\mathbf{t}}^1 = h_3 h_1 (\nabla u^3 \times \nabla u^1), \\ \hat{\mathbf{t}}^3 &= \hat{\mathbf{t}}^1 \times \hat{\mathbf{t}}^2 = h_1 h_2 (\nabla u^1 \times \nabla u^2); \end{aligned}\right\} \qquad (2.6)$$

$\mathbf{F}$ is expressed as

$$\mathbf{F}(u^{\alpha}) \equiv \sum (f_1\, h_2\, h_3)(\nabla u^2 \times \nabla u^3), \tag{2.7}$$

where $\sum$ stands for the sum of two additional terms obtained by cyclic interchange of the indices 1, 2, 3. Operating Eq. (2.7) by $\nabla.$ and simplifying by means of

$$\nabla \cdot (f\mathbf{a}) = (\nabla f) \cdot \mathbf{a} + f(\nabla \cdot \mathbf{a}), \tag{2.8}$$

and

$$\nabla \cdot (\nabla u^2 \times \nabla u^3) = (\nabla \times \nabla u^2) \cdot \nabla u^3 - (\nabla u^2) \cdot (\nabla \times \nabla u^3) = 0, \tag{2.9}$$

we get

$$\nabla \cdot \mathbf{F} = \Sigma\{\, \nabla (f_1\, h_2\, h_3)\, \} \cdot (\nabla u^2 \times \nabla u^3)$$

$$= \Sigma\, \{(f_1\, h_2\, h_3)_{,\alpha}\, \nabla u^{\alpha}\, \} \cdot (\nabla u^2 \times \nabla u^3), \qquad \text{by Eq. (2.3a)}$$

$$= \{\Sigma\, (f_1\, h_2\, h_3)_{,1}\, \}\, [\, \nabla u^1\ \ \nabla u^2\ \ \nabla u^3] = \{\, \Sigma\, (f_1\, h_2\, h_3)_{,1}\, \} \,/\, K,$$

by Eq. (1.20). //

Corollary 2.1. The Laplacian of a scalar point function f is

$$\nabla^2 f \equiv \nabla \cdot (\nabla f) = \{\Sigma\, (f_{,1}\, h_2\, h_3 / h_1)_{,1}\, \} \,/\, K. \tag{2.10}$$

Proof. Comparing Eqs. (2.3b) and (2.4) the vector function $\mathbf{F} \equiv \nabla f$ has components

$$f_{\beta} = f_{,\beta} / h_{\beta} \tag{2.11}$$

Along the orthonormal vectors $\hat{\mathbf{t}}^{\beta}$. Accordingly, Eq. (2.5) assumes the form of Eq. (2.10). //

Theorem 2.2. The curl of the vector function defined by Eq. (2.4) is given by

$$\nabla \times \mathbf{F} = \sum \{(f_3\ h_3)_{,2} - (f_2\ h_2)_{,3}\}\ \hat{\mathbf{t}}^1 / h_2 h_3. \tag{2.12}$$

Proof. Putting for $\hat{\mathbf{t}}^{\beta}$ from Eq. (1.19), the function $\mathbf{F}$ given by Eq. (2.4) is rewritten as

$$\mathbf{F} = f_{\beta} h_{\beta}\, \nabla u^{\beta} \equiv f_1 h_1 \nabla u^1 + f_2 h_2 \nabla u^2 + f_3 h_3 \nabla u^3. \tag{2.13}$$

Operating it by $\nabla\times$ and using

$$\nabla \times (f\mathbf{a}) = (\nabla f) \times \mathbf{a} + f \nabla \times \mathbf{a}, \tag{2.14}$$

and

$$\nabla \times (\nabla u) = 0, \tag{2.15}$$

we get

$$\nabla\times\mathbf{F} = \{\nabla(f_\beta h_\beta)\}\times\nabla u^\beta = \{(f_\beta h_\beta)_{,\alpha}\nabla u^\alpha\}\times\nabla u^\beta \qquad \text{by Eq. (2.3a)}$$

$$= \sum\{(f_1 h_1)_{,2}(\nabla u^2\times\nabla u^1)+(f_1 h_1)_{,3}(\nabla u^3\times\nabla u^1)\}$$

$$= -(f_1 h_1)_{,2}\,\hat{\mathbf{t}}^3/h_1h_2 + (f_1 h_1)_{,3}\,\hat{\mathbf{t}}^2/h_3h_1 - (f_2 h_2)_{,3}\,\hat{\mathbf{t}}^1/h_2h_3$$

$$+(f_2 h_2)_{,1}\,\hat{\mathbf{t}}^3/h_1h_2 - (f_3 h_3)_{,1}\,\hat{\mathbf{t}}^2/h_3h_1 + (f_3 h_3)_{,2}\,\hat{\mathbf{t}}^1/h_2h_3,$$

by Eq. (2.6). Thus, we get Eq. (2.12). //

§ 3. Cylindrical coordinates

Reverting back to non-index notation let PIJKLMNQ be a cylinder and P (x, y, z) be a point on it with its projection I (x, y) on the xOy - plane. Let OI = ρ make angle φ with x-axis then ρ, φ, z are connected, with the rectangular Cartesian coordinates x, y, z, by

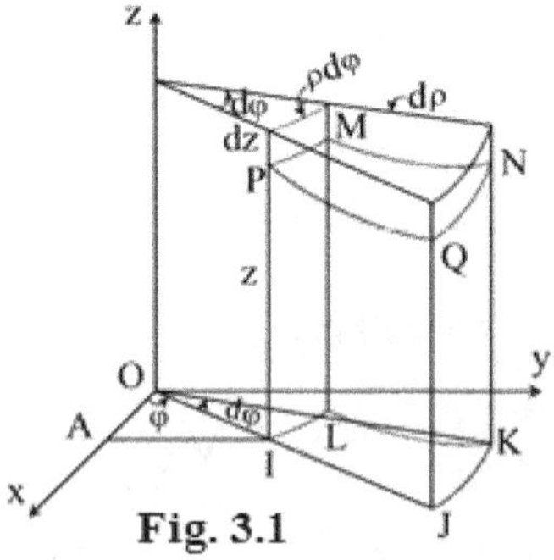

Fig. 3.1

$$x = \rho\cos\varphi, \quad y = \rho\sin\varphi, \quad z = z; \tag{3.1a}$$

and are called the *cylindrical coordinates* of the point P. The level surfaces $\rho = \rho_o$, $\varphi = \varphi_o$ and $z = z_o$ are respectively cylinders about the z-axis, planes through and planes perpendicular to the z-axis. The coordinate curves for ρ are rays perpendicular to z-axis; for φ, the horizontal circles with centers on z-axis and for z, the lines parallel to the z-axis.

The position vector $\mathbf{r}$ of P as given by Eq. (1.1) becomes

$$\mathbf{r} = x\hat{\mathbf{i}} + y\hat{\mathbf{j}} + z\hat{\mathbf{k}} = (\rho\cos\varphi,\ \rho\sin\varphi,\ z), \tag{3.1b}$$

where $\hat{\mathbf{i}}$, $\hat{\mathbf{j}}$, $\hat{\mathbf{k}}$ are the unit vectors along the respective coordinate axes Ox, Oy, Oz. The differentials of Eq. (3.1a)

$$\left.\begin{aligned} dx &= (\cos\varphi)\,d\rho - (\rho\sin\varphi)\,d\varphi, \\ dy &= (\sin\varphi)\,d\rho + (\rho\cos\varphi)\,d\varphi, \qquad dz = dz \end{aligned}\right\} \tag{3.2}$$

determine the infinitesimal vector $\overrightarrow{PQ} = d\mathbf{r}$. Therefore, the infinitesimal arc-length $PQ = ds$ is given by

$$(ds)^2 = d\mathbf{r} \,.\, d\mathbf{r} = (dx)^2 + (dy)^2 + (dz)^2 \tag{3.3a}$$

$$= (d\rho)^2 + \rho^2 (d\varphi)^2 + (dz)^2. \tag{3.3b}$$

Also, derivations of Eq. (3.1b) yield the vectors

$$\left.\begin{aligned} \hat{\mathbf{t}}_\rho &\equiv \partial \mathbf{r}/\partial \rho = (\cos\varphi, \sin\varphi, 0), \\ \rho\hat{\mathbf{t}}_\varphi &\equiv \partial \mathbf{r}/\partial\varphi = (-\rho\sin\varphi, \rho\cos\varphi, 0), \\ \text{and} \quad \hat{\mathbf{t}}_z &\equiv \partial \mathbf{r}/\partial z = (0, 0, 1), \end{aligned}\right\} \tag{3.4}$$

with their magnitudes:

$$h_1 \equiv |\, \partial \mathbf{r}/\partial\rho \,| = 1, \quad h_2 \equiv |\, \partial \mathbf{r}/\partial\varphi \,| = \rho, \quad h_3 \equiv |\, \partial \mathbf{r}/\partial z \,| = 1. \tag{3.5}$$

Hence, Eq. (1.17) determines $K = \rho$. The vectors $\hat{\mathbf{t}}_\rho$, $\hat{\mathbf{t}}_\varphi$ and $\hat{\mathbf{t}}_z$ are the unit tangent vectors to the respective ρ-, φ- and z-curves.

Theorem 3.1. The cylindrical coordinate system (ρ, φ, z) is orthogonal and right-handed.

Proof. In view of Eq. (3.4), the vectors $\hat{\mathbf{t}}_\rho$, $\hat{\mathbf{t}}_\varphi$ and $\hat{\mathbf{t}}_z$ satisfy

$$\hat{\mathbf{t}}_\rho \,.\, \hat{\mathbf{t}}_\varphi = \hat{\mathbf{t}}_\varphi \,.\, \hat{\mathbf{t}}_z = \hat{\mathbf{t}}_z \,.\, \hat{\mathbf{t}}_\rho = 0. \tag{3.6}$$

Thus, orthogonality of the system is established. Further,

$$\hat{\mathbf{t}}_\rho \times \hat{\mathbf{t}}_\varphi = \hat{\mathbf{t}}_z, \quad \hat{\mathbf{t}}_\varphi \times \hat{\mathbf{t}}_z = \hat{\mathbf{t}}_\rho, \quad \hat{\mathbf{t}}_z \times \hat{\mathbf{t}}_\rho = \hat{\mathbf{t}}_\varphi \tag{3.7}$$

make the system right-handed. //

Above theorem justifies the metric tensor components

$$g_{11} = g_{33} = 1, \quad g_{22} = \rho^2, \quad g_{\alpha\beta} = 0 \quad \text{when } \alpha \neq \beta; \tag{3.8}$$

which can be derived from Eq. (3.3b). These components span the matrix

$$((g_{\alpha\beta})) = \begin{bmatrix} 1 & 0 & 0 \\ 0 & \rho^2 & 0 \\ 0 & 0 & 1 \end{bmatrix} \text{(3.9); with determinant } g = |g_{\alpha\beta}| = \rho^2. \quad (3.10)$$

Hence, the associate metric tensor with respect to the cylindrical coordinates can be found from [12], Eq. (15.5.1):

$$g^{11} = g^{33} = 1, \quad g^{22} = 1/\rho^2, \quad g^{\alpha\beta} = 0, \text{ when } \alpha \neq \beta; \quad (3.11a)$$

spanning the matrix

$$((g^{\alpha\beta})) = \begin{bmatrix} 1 & 0 & 0 \\ 0 & 1/\rho^2 & 0 \\ 0 & 0 & 1 \end{bmatrix}. \quad (3.11b)$$

However, in the following we also derive the components of the metric tensor by coordinate transformation vide Eq. (3.1a) of the rectangular Cartesian coordinates x, y, z onto the cylindrical coordinates ρ, φ, z.

Theorem 3.2. The metric tensor components

$$g_{ij} = \delta_{ij} \quad (3.12)$$

with respect to the rectangular Cartesian coordinates transform under the coordinate transformation in Eq. (3.1a) onto the components as given by Eq. (3.8).

Proof. Switching over to index notation let the rectangular coordinates x, y, z be denoted by x^i's and the cylindrical coordinates ρ, φ, z by $\bar{x}^i$'s respectively. Therefore, from Eq. (3.1a), the projection factors are:

$$\left.\begin{array}{ll} \partial x^1/\partial \bar{x}^1 \equiv \partial x/\partial\rho = \cos\varphi, & \partial x^1/\partial \bar{x}^2 \equiv \partial x/\partial\varphi = -\rho\sin\varphi, \\ \partial x^1/\partial \bar{x}^3 \equiv \partial x/\partial z = 0; & \partial x^2/\partial \bar{x}^1 \equiv \partial y/\partial\rho = \sin\varphi, \\ \partial x^2/\partial \bar{x}^2 \equiv \partial y/\partial\varphi = \rho\cos\varphi, & \partial x^2/\partial \bar{x}^3 \equiv \partial y/\partial z = 0; \\ \partial x^3/\partial \bar{x}^1 \equiv \partial z/\partial\rho = 0, & \partial x^3/\partial \bar{x}^2 \equiv \partial z/\partial\varphi = 0, \\ & \partial x^3/\partial \bar{x}^3 \equiv \partial z/\partial z = 1. \end{array}\right\} \quad (3.13)$$

Rewriting the coordinate transformation law in [12], Eq. (15.4.3a) of the metric tensor when Eq. (3.12) holds:

$$\bar{g}_{\alpha\beta} = (\partial x^1/\partial \bar{x}^\alpha)(\partial x^1/\partial \bar{x}^\beta) + (\partial x^2/\partial \bar{x}^\alpha)(\partial x^2/\partial \bar{x}^\beta) + (\partial x^3/\partial \bar{x}^\alpha)(\partial x^3/\partial \bar{x}^\beta);$$

and putting from Eq. (3.13), we obtain

$$\bar{g}_{11} = (\partial x^1/\partial \bar{x}^1)^2 + (\partial x^2/\partial \bar{x}^1)^2 + (\partial x^3/\partial \bar{x}^1)^2 = 1$$

$$\bar{g}_{22} = (\partial x^1/\partial \bar{x}^2)^2 + (\partial x^2/\partial \bar{x}^2)^2 + (\partial x^3/\partial \bar{x}^2)^2 = \rho^2,$$

$$\bar{g}_{33} = (\partial x^1/\partial \bar{x}^3)^2 + (\partial x^2/\partial \bar{x}^3)^2 + (\partial x^3/\partial \bar{x}^3)^2 = 1,$$

and

$$\bar{g}_{\alpha\beta} = 0 \quad \text{when } \alpha \neq \beta. \; //$$

Theorem 3.3. The only non-zero components of the Christoffel symbols of first kind in the cylindrical coordinates are

$$[12, 2] = [21, 2] = \rho \qquad (3.14); \qquad [22, 1] = -\rho. \quad (3.15)$$

Proof. Differentiating Eq. (3.8) partially with respect to ρ, φ, z we note that the only non-vanishing derivatives of $g_{\alpha\beta}$ are

$$g_{22,1} \equiv \partial g_{22}/\partial \rho = 2\rho. \qquad (3.16)$$

Accordingly, Eqs. (15.7.1) of [12] and (3.8) yield

$$[12, 2] = [21, 2] = (1/2)\, g_{22,1} \text{ and } [22, 1] = -(1/2)\, g_{22,1};$$

which, for Eq. (3.16), assume the forms of Eqs. (3.14) and (3.15) above. Rest all the 24 components of $[\alpha\, \beta, \gamma\,]$ can be seen zero. //

Theorem 3.4. The second kind Christoffel symbols in the coordinates (ρ, φ, z) are given by

$$\left\{ {1 \atop 2\,2} \right\} = -\rho, \; \left\{ {2 \atop 1\,2} \right\} = \left\{ {2 \atop 2\,1} \right\} = 1/\rho, \qquad (3.17)$$

and rest are zero.

Proof. Applying Eq. (15.7.2) of [12] and putting from Eq. (3.11a), we find

$$\left\{ {1 \atop 2\ 2} \right\} = g^{1\alpha}[22, \alpha] = g^{11}\ [22, 1],$$

and

$$\left\{ {2 \atop 1\ 2} \right\} = \left\{ {2 \atop 2\ 1} \right\} = g^{2\alpha}[12, \alpha] = g^{22}\ [12, 2];$$

which, for Eqs. (3.11a), (3.14) and (3.15), reduce to the forms in Eqs. (3.17). Rest of the components of $\left\{ {\alpha \atop \beta\ \gamma} \right\}$ vanish identically for similar reasons. //

Theorem 3.5. The Riemann-Christoffel symbols of either kind vanish identically in the cylindrical coordinate system.

Proof. As seen in [12], Example 15.14.1, the associate curvature tensor may have at most the following independent components:

$$R_{1212},\quad R_{1213},\quad R_{1223},\quad R_{1313},\quad R_{1323}\quad \text{and}\quad R_{2323}.$$

Writing their values according to Eq. (15.14.7) of [12], and employing the results of Theorem 3.3 all of them can be seen zero. Further, it follows from Eq. (15.14.10) of [12], that the curvature tensor too vanishes identically. //

Example 3.1. A covariant vector has components

$$A_1 = 2x - z, \quad A_2 = x^2 y, \quad A_3 = yz \tag{3.18}$$

in the rectangular Cartesian coordinate system. Find its components in the cylindrical coordinates.

Solution. The projection factors $\partial x^i / \partial \bar{x}^\alpha$ of the coordinate transformation in Eq. (3.1a) are given by Eq. (3.13). Therefore, the components $\bar{A}_\alpha$ of the said vector with respect to the coordinates ρ, φ, z transform according to the coordinate transformation law vide Eq. (4.5.1b):

$$\bar{A}_\alpha = A_i(\partial x^i / \partial \bar{x}^\alpha)$$

$$= A_1(\partial x^1 / \partial \bar{x}^\alpha) + A_2(\partial x^2 / \partial \bar{x}^\alpha) + A_3(\partial x^3 / \partial \bar{x}^\alpha). \tag{3.19}$$

Putting from Eqs. (3.1a), (3.13) and (3.18), we obtain

$$\overline{A}_1 = (2x - z)\cos\varphi + x^2 y \sin\varphi$$

$$= (2\rho\cos\varphi - z)\cos\varphi + \rho^3\cos^2\varphi\sin^2\varphi, \tag{3.20}$$

$$\overline{A}_2 = (2x - z)(-\rho.\sin\varphi) + x^2 y\rho.\cos\varphi$$

$$= \rho\sin\varphi.(z - 2\rho\cos\varphi + \rho^3\cos^3\varphi), \tag{3.21}$$

and

$$\overline{A}_3 = yz = \rho z\sin\varphi. \text{ //} \tag{3.22}$$

§ 4. ∇ operator in cylindrical coordinates

4.1. ∇*f*. In view of Eq. (3.5), the expression for ∇f given by Eq. (2.3b), for cylindrical coordinates, simplifies to

$$\nabla f = (\partial f/\partial\rho)\hat{\mathbf{t}}_\rho + (1/\rho)(\partial f/\partial\varphi)\hat{\mathbf{t}}_\varphi + (\partial f/\partial z)\hat{\mathbf{t}}_z, \tag{4.1}$$

or, by Eq. (3.4),

$$\nabla f = \left(\cos\varphi\cdot\frac{\partial f}{\partial\rho} - \frac{\sin\varphi}{\rho}\cdot\frac{\partial f}{\partial\varphi},\ \sin\varphi\cdot\frac{\partial f}{\partial\rho} + \frac{\cos\varphi}{\rho}\cdot\frac{\partial f}{\partial\varphi},\ \frac{\partial f}{\partial z}\right). \tag{4.2}$$

4.2. ∇ . F. The divergence of a vector point function

$$\mathbf{F}(\rho, \varphi, z) \equiv f_1\,\hat{\mathbf{t}}_\rho + f_2\,\hat{\mathbf{t}}_\varphi + f_3\,\hat{\mathbf{t}}_z \tag{4.3}$$

of cylindrical coordinates is obtained from Eqs. (2.5) and (3.5):

$$\nabla.\mathbf{F} = (1/\rho)\{\partial(\rho f_1)/\partial\rho + \partial f_2/\partial\varphi + \partial(\rho f_3)/\partial z\}$$

$$= (1/\rho)(f_1 + \partial f_2/\partial\varphi) + \partial f_1/\partial\rho + \partial f_3/\partial z. \tag{4.4}$$

4.3. ∇² *f*. Putting from Eq. (3.5) in Eq. (2.10), the Laplacian of a scalar function f in cylindrical coordinates is obtained by

$$\nabla^2 f = \frac{1}{\rho}\left\{\frac{\partial}{\partial\rho}\left(\frac{\rho\,\partial f}{\partial\rho}\right) + \frac{\partial}{\partial\varphi}\left(\frac{1}{\rho}\cdot\frac{\partial f}{\partial\varphi}\right) + \frac{\partial}{\partial z}\left(\frac{\rho\,\partial f}{\partial z}\right)\right\}$$

$$= \frac{1}{\rho}\cdot\frac{\partial f}{\partial\rho}+\frac{\partial^2 f}{\partial\rho^2}+\frac{1}{\rho^2}\cdot\frac{\partial^2 f}{\partial\varphi^2}+\frac{\partial^2 f}{\partial z^2}. \tag{4.5}$$

4.4. ∇ × F. The curl of a vector point function is given by Eq. (2.12). The same for the function given by in Eq. (4.3) of cylindrical coordinates satisfying Eq. (3.5) reduces to

$$\nabla\times\mathbf{F} = \frac{1}{\rho}\left\{\frac{\partial f_3}{\partial\varphi}-\frac{\partial(\rho f_2)}{\partial z}\right\}\hat{\mathbf{t}}_\rho+\left(\frac{\partial f_1}{\partial z}-\frac{\partial f_3}{\partial\rho}\right)\hat{\mathbf{t}}_\varphi+\frac{1}{\rho}\left\{\frac{\partial(\rho f_2)}{\partial\rho}-\frac{\partial f_1}{\partial\varphi}\right\}\hat{\mathbf{t}}_z$$

$$=\left(\frac{1}{\rho}\cdot\frac{\partial f_3}{\partial\varphi}-\frac{\partial f_2}{\partial z}\right)\hat{\mathbf{t}}_\rho+\left(\frac{\partial f_1}{\partial z}-\frac{\partial f_3}{\partial\rho}\right)\hat{\mathbf{t}}_\varphi+\left\{\frac{1}{\rho}\left(f_2-\frac{\partial f_1}{\partial\varphi}\right)+\frac{\partial f_2}{\partial\rho}\right\}\hat{\mathbf{t}}_z, \tag{4.6}$$

or, equivalently,

$$\nabla\times\mathbf{F} = \begin{vmatrix}\hat{\mathbf{t}}_\rho/\rho & \hat{\mathbf{t}}_\varphi & \hat{\mathbf{t}}_z/\rho\\ \partial/\partial\rho & \partial/\partial\varphi & \partial/\partial z\\ f_1 & \rho f_2 & f_3\end{vmatrix}. \tag{4.7}$$

Example 4.1. Both $\nabla(\ln\rho)$ and $\nabla\varphi$ are solenoidal vectors.

Solution. The function $f = \ln\rho$ being of variable ρ only has its derivatives with respect to cylindrical coordinates:

$$\partial f/\partial\rho = 1/\rho,\quad \partial^2 f/\partial\rho^2 = -1/\rho^2,\quad \partial f/\partial\varphi = \partial f/\partial z = 0;$$

which make the divergence of $\nabla\rho$, given by Eq. (4.5), zero.

Also, the divergence of $\nabla\varphi$, i.e. Laplacian of φ, is seen term-wise zero when f is replaced by φ in Eq. (4.5). Hence, the statement follows in view of Definition 2.3.1. //

Example 4.2. The vector function

$$\mathbf{F} = \rho z\sin 2\varphi\,(\hat{\mathbf{t}}_\rho+\hat{\mathbf{t}}_\varphi\cot 2\varphi+\rho\hat{\mathbf{t}}_z/2z) \tag{4.8}$$

is irrotational.

Solution. Comparing Eqs. (4.3) and (4.8), we have

$$f_1 = \rho z\sin 2\varphi,\quad f_2 = \rho z\cos 2\varphi,\quad f_3 = (\rho^2/2)\sin 2\varphi;$$

which possess their derivatives

$$\partial f_1 / \partial \rho = z \sin 2\varphi, \quad \partial f_1 / \partial \varphi = 2\rho z \cos 2\varphi, \quad \partial f_1 / \partial z = \rho \sin 2\varphi;$$

$$\partial f_2 / \partial \rho = z \cos 2\varphi, \quad \partial f_2 / \partial \varphi = -2\rho z \sin 2\varphi, \quad \partial f_2 / \partial z = \rho \cos 2\varphi;$$

$$\partial f_3 / \partial \rho = \rho \sin 2\varphi, \quad \partial f_3 / \partial \varphi = \rho^2 \cos 2\varphi, \quad \partial f_3 / \partial z = 0.$$

Putting for these derivatives in Eq. (4.6), the curl of **F** is seen as a null vector. Hence, in view of Definition 13.4.1, **F** becomes irrotational. //

§ 5. Spherical polar coordinates

Let P (x, y, z) be a point in the space E_3 with its projection I (x,y) on the xOy-plane. Let OP $= r$ be inclined at an angle θ to the z-axis and OI $= r \sin\theta$ make angle φ with the x-axis. The projections OA $= x$, OB $=$ AI $= y$ and OC $=$ IP $= z$ of OP on the rectangular Cartesian coordinate axes Ox, Oy, Oz are given by

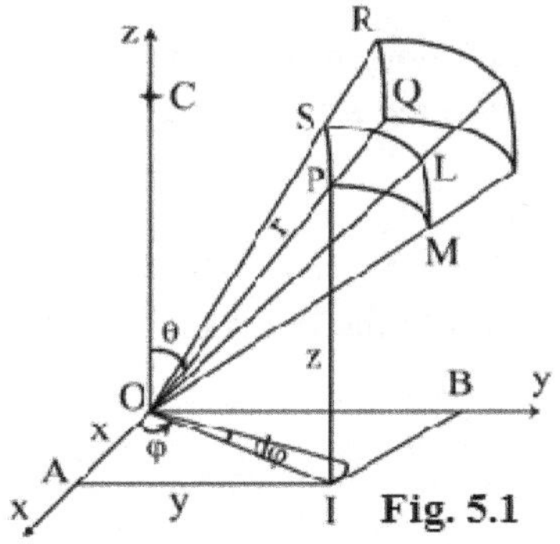

Fig. 5.1

$$x = (\text{OI}) \cos\varphi = r \sin\theta \cos\varphi, \quad y = r \sin\theta \sin\varphi, \quad z = r\cos\theta. \qquad (5.1)$$

Accordingly, the position vector **r** of P with respect to origin O and above coordinate axes is given by

$$\mathbf{r} = x\,\hat{\mathbf{i}} + y\,\hat{\mathbf{j}} + z\,\hat{\mathbf{k}} = r\,(\sin\theta \cos\varphi,\ \sin\theta \sin\varphi,\ \cos\theta). \qquad (5.2)$$

The variables r, θ, φ are called the *spherical polar coordinates* of the point P.

The level surfaces $r = r_o$, $\theta = \theta_o$ and $\varphi = \varphi_o$ are respectively spheres about O, cones about z-axis with vertex at O, and planes through the z-axis. The coordinate curves are

(i) rays from the origin for r-curves,

(ii) vertical circles centred at O (called the *meridians*) for θ,

(iii) horizontal circles with their centres on the z-axis for φ.

The differentials of Eq. (5.1):

$$\left.\begin{aligned} dx &= (dr)\sin\theta\cos\varphi + r\cos\theta\cos\varphi\, d\theta - r\sin\theta\sin\varphi\, d\varphi), \\ dy &= (dr)\sin\theta\sin\varphi + r\cos\theta\sin\varphi\, d\theta + r\sin\theta\cos\varphi\, d\varphi), \\ dz &= (dr)\cos\theta - r\sin\theta\, d\theta\,. \end{aligned}\right\} \quad (5.3)$$

determine the infinitesimal arc-length ds:

$$(ds)^2 = (dx)^2 + (dy)^2 + (dz)^2 = (dr)^2 + (rd\theta)^2 + (r\sin\theta\, d\varphi)^2; \quad (5.4)$$

which determines the metric of the space in spherical coordinates. The components of the metric tensor are

$$g_{11} = 1, \quad g_{22} = r^2, \quad g_{33} = r^2\sin^2\theta, \quad g_{\alpha\beta} = 0 \text{ when } \alpha \neq \beta; \quad (5.5a)$$

spanning the matrix

$$((g_{\alpha\beta})) = \begin{bmatrix} 1 & 0 & 0 \\ 0 & r^2 & 0 \\ 0 & 0 & r^2\sin^2\theta \end{bmatrix}, \quad (5.5b)$$

with its determinant

$$g = r^4\sin^2\theta. \quad (5.6)$$

The associate metric tensor has components

$$g^{11} = 1, \quad g^{22} = 1/r^2, \quad g^{33} = 1/r^2\sin^2\theta, \quad g^{\alpha\beta} = 0, \text{ when } \alpha \neq \beta. \quad (5.7)$$

The tangent vectors to the coordinate curves are obtained by differentiating Eq. (5.2) with respect to the parameters r, θ, φ:

$$\left.\begin{aligned} \partial\mathbf{r}/\partial r &= (\sin\theta\cos\varphi, \sin\theta\sin\varphi, \cos\theta) = \hat{\mathbf{t}}_r, \\ \partial\mathbf{r}/\partial\theta &= r(\cos\theta\cos\varphi, \cos\theta\sin\varphi, -\sin\theta) = r\hat{\mathbf{t}}_\theta, \\ \partial\mathbf{r}/\partial\varphi &= r\sin\theta\,(-\sin\varphi, \cos\varphi, 0) = (r\sin\theta)\ \hat{\mathbf{t}}_\varphi; \end{aligned}\right\} \quad (5.8)$$

where $\hat{\mathbf{t}}_r$, $\hat{\mathbf{t}}_\theta$ and $\hat{\mathbf{t}}_\varphi$ are the unit vectors along the respective tangents. Comparing Eqs. (5.8) and (1.15), we find

$$h_1 \equiv |\partial \mathbf{r} / \partial r| = 1, \quad h_2 \equiv |\partial \mathbf{r} / \partial \theta| = r, \quad h_3 \equiv |\partial \mathbf{r} / \partial \varphi| = r \sin \theta. \tag{5.9}$$

Theorem 5.1. The spherical polar coordinate system is orthogonal and the vectors $\hat{\mathbf{t}}_r$, $\hat{\mathbf{t}}_\theta$, $\hat{\mathbf{t}}_\varphi$ are right-handed.

Proof. Forming scalar products of the vectors $\hat{\mathbf{t}}_r$, $\hat{\mathbf{t}}_\theta$ and $\hat{\mathbf{t}}_\varphi$ we get

$$\hat{\mathbf{t}}_r . \hat{\mathbf{t}}_\theta = \hat{\mathbf{t}}_\theta . \hat{\mathbf{t}}_\varphi = \hat{\mathbf{t}}_\varphi . \hat{\mathbf{t}}_r = 0; \tag{5.10}$$

which establishes their orthogonality. Further, their cross products satisfy

$$\hat{\mathbf{t}}_r \times \hat{\mathbf{t}}_\theta = \begin{vmatrix} \hat{\mathbf{i}} & \hat{\mathbf{j}} & \hat{\mathbf{k}} \\ \sin\theta \cos\varphi & \sin\theta \sin\varphi & \cos\theta \\ \cos\theta \cos\varphi & \cos\theta \sin\varphi & -\sin\theta \end{vmatrix} = \hat{\mathbf{t}}_\varphi, \tag{5.11a}$$

$$\hat{\mathbf{t}}_\theta \times \hat{\mathbf{t}}_\varphi = \begin{vmatrix} \hat{\mathbf{i}} & \hat{\mathbf{j}} & \hat{\mathbf{k}} \\ \cos\theta \cos\varphi & \cos\theta \sin\varphi & -\sin\theta \\ -\sin\varphi & \cos\varphi & 0 \end{vmatrix} = \hat{\mathbf{t}}_r, \tag{5.11b}$$

$$\hat{\mathbf{t}}_\varphi \times \hat{\mathbf{t}}_r = \begin{vmatrix} \hat{\mathbf{i}} & \hat{\mathbf{j}} & \hat{\mathbf{k}} \\ -\sin\varphi & \cos\varphi & 0 \\ \sin\theta \cos\varphi & \sin\theta \sin\varphi & \cos\theta \end{vmatrix} = \hat{\mathbf{t}}_\theta \cdot \tag{5.11c}$$

So, they form a right-handed system of orthonormal vectors. //

Theorem 5.2. The only non-vanishing components of the Christoffel symbols of the first kind in the spherical coordinates are

$$\left.\begin{aligned} [1\,2, 2] &= [2\,1, 2] = -[2\,2, 1] = r, \\ [1\,3, 3] &= [3\,1, 3] = -[3\,3, 1] = r \sin^2\theta, \\ [2\,3, 3] &= [3\,2, 3] = -[3\,3, 2] = r^2 \sin\theta \cos\theta. \end{aligned}\right\} \tag{5.12}$$

Proof. The only non-zero derivatives of the components of the metric tensor, obtainable from Eq. (5.5a), are

$$\left.\begin{aligned} g_{22,1} &\equiv \partial r^2 / \partial r = 2r, \qquad g_{33,1} \equiv \partial (r\sin\theta)^2 / \partial r = 2r\ \sin^2\theta, \\ g_{33,2} &\equiv \partial (r\sin\theta)^2 / \partial\theta = 2r^2\ \sin\theta\ \ \cos\theta. \end{aligned}\right\} \tag{5.13}$$

Consequently, Eq. (15.7.1) of [12] yields Eq. (5.12). //

Theorem 5.3. The non-zero components of the second kind Christoffel symbols in the coordinates r, θ, φ are

$$\left.\begin{aligned} &\left\{ {1 \atop 2\ 2} \right\} = -r, \left\{ {1 \atop 3\ 3} \right\} = -r\sin^2\theta, \left\{ {2 \atop 3\ 3} \right\} = -\sin\theta\ \cos\theta, \\ &\left\{ {2 \atop 1\ 2} \right\} = \left\{ {2 \atop 2\ 1} \right\} = \left\{ {3 \atop 1\ 3} \right\} \equiv \left\{ {3 \atop 3\ 1} \right\} = 1/r, \ \left\{ {3 \atop 2\ 3} \right\} \equiv \left\{ {3 \atop 3\ 2} \right\} = \cot\theta. \end{aligned}\right\} \tag{5.14}$$

Proof. Expanding Eq. (15.7.2) of [12] and putting from Eq. (5.7), we obtain

$$\left\{ {1 \atop 2\ 2} \right\} = g^{1\alpha}[2\ 2,\ \alpha] = [2\ 2,\ 1], \ \left\{ {1 \atop 3\ 3} \right\} = g^{1\alpha}[3\ 3,\ \alpha] = [3\ 3,\ 1],$$

$$\left\{ {2 \atop 1\ 2} \right\} = \left\{ {2 \atop 2\ 1} \right\} = g^{2\alpha}[1\ 2,\ \alpha] = [1\ 2,\ 2]/r^2,$$

$$\left\{ {2 \atop 3\ 3} \right\} = g^{2\alpha}[3\ 3,\ \alpha] = [3\ 3,\ 2]/r^2,$$

$$\left\{ {3 \atop 1\ 3} \right\} = \left\{ {3 \atop 3\ 1} \right\} = g^{3\alpha}[1\ 3,\ \alpha] = [1\ 3,\ 3]/r^2\sin^2\theta,$$

$$\left\{ {3 \atop 2\ 3} \right\} = \left\{ {3 \atop 3\ 2} \right\} = g^{3\alpha}[2\ 3,\ \alpha] = [2\ 3,\ 3]/r^2\sin^2\theta\,;$$

which, for Eqs. (5.12), assume the form of Eqs. (5.14). //

Theorem 5.4. In the spherical polar coordinates both the curvature tensor and the associate curvature tensor vanish identically.

Proof. As seen in [12], Example 15.14.1 and Theo. 3.5, there may exist at most six independent components (given therein) of the associate curvature tensor. Writing their expressions according to Eq. (15.14.7) of [12] and using Theo. 5.2, the components R_{1213} and R_{1223} are termwise zero; whereas

$$R_{1212} = \partial_1 [2\,1, 2] - g^{22} [1\,2, 2]^2 = (\partial r/\partial r) - r^2 / r^2,$$

$$R_{1313} = \partial_1 [31, 3] - g^{33} [13, 3]^2 = \partial (r \sin^2 \theta)/\partial r - (r \sin^2 \theta / r \sin \theta)^2,$$

$$R_{1323} = \partial_1 [32, 3] + g^{22} [12, 2] [33, 2] - g^{33} [32, 3] [13, 3]$$

$$= \partial (r^2 \sin \theta \cos \theta) / \partial r - (r \,.\, r^2 \sin \theta \cos \theta) / r^2$$

$$- (r^2 \sin \theta \cos \theta). (r \sin^2 \theta) / (r \sin \theta)^2,$$

and

$$R_{2323} = \partial_2 [32, 3] + g^{11} [22, 1] [33, 1] - g^{33} [23, 3]^2$$

$$= \partial (r^2 \sin \theta \cos \theta) / \partial \theta + r.\, r \sin^2 \theta - (r^2 \sin \theta \cos \theta)^2 / (r \sin \theta)^2,$$

also vanish for Eqs. (5.7) and (5.12). Thus, all the components of the associate curvature tensor are zero. Further, vanishing property of the curvature tensor follows from Eq. (15.14.10) of [12] and above discussion. //

Example 5.1. Find the components of the metric tensor and of its associate tensor for the metric

$$(ds)^2 = (adr)^2 / (a^2 - r^2) + r^2 \{(d\theta)^2 + \sin^2 \theta \, (d\varphi)^2\}. \tag{5.15}$$

Solution. A comparison of Eqs. (5.4) and (5.15) determines the components of the metric tensor:

$$g_{11} = a^2 / (a^2 - r^2), \quad g_{22} = r^2, \quad g_{33} = r^2 \sin^2 \theta, \quad g_{\alpha\beta} = 0, \tag{5.16}$$

when $\alpha \neq \beta$; so that

$$g = \begin{vmatrix} g_{11} & 0 & 0 \\ 0 & g_{22} & 0 \\ 0 & 0 & g_{33} \end{vmatrix} = g_{11} g_{22} g_{33} = \frac{(a\, r^2 \sin \theta)^2}{a^2 - r^2}. \tag{5.17}$$

The components of the associate metric tensor are given by [12], Eq. (15.5.1):

$$\left.\begin{aligned} g^{11} = 1 / g_{11} = (a^2 - r^2) / a^2, \quad g^{22} = 1/ g_{22} = 1/ r^2, \\ g^{33} = 1 / g_{33} = 1 / (r \sin \theta)^2, \quad g^{\alpha\beta} = 0 \ \text{ when } \alpha \neq \beta. \ // \end{aligned}\right\} \tag{5.18}$$

Example 5.2. A contravariant vector has components a, b, c in rectangular Cartesian coordinates. Find its components in the spherical polar coordinates.

Solution. Solving Eqs. (5.1), we get their inverse transformation:

$$r^2 = x^2 + y^2 + z^2, \quad \theta = \cos^{-1}(z/r), \quad \varphi = \tan^{-1}(y/x). \tag{5.19}$$

Employing the index notation:

$$x^1 = x, \quad x^2 = y, \quad x^3 = z \text{ and } \bar{x}^1 = r, \quad \bar{x}^2 = \theta, \quad \bar{x}^3 = \varphi; \tag{5.20}$$

and differentiating Eqs. (5.19), we derive the projection factors:

$$\partial \bar{x}^1/\partial x^1 \equiv \partial r / \partial x = x/r = \sin\theta \cos\varphi,$$

$$\partial \bar{x}^1/ \partial x^2 \equiv \partial r / \partial y = y / r = \sin\theta . \sin\varphi,$$

$$\partial \bar{x}^1/ \partial x^3 \equiv \partial r / \partial z = z / r = \cos\theta;$$

$$\partial \bar{x}^2/ \partial x^1 \equiv \partial\theta / \partial x = \{-1/\surd(1 - z^2/r^2)\}(-z/r^2)(x/r)$$

$$= z\,x / r^2 \surd(r^2 - z^2) = (\cos\theta \cos\varphi)/r,$$

$$\partial \bar{x}^2/ \partial x^2 \equiv \partial\theta / \partial y = \{-1/\surd(1 - z^2/r^2)\}(-z/r^2)(y/r)$$

$$= yz / r^2 \surd(r^2 - z^2) = (\cos\theta \sin\varphi)/r,$$

$$\partial \bar{x}^2/ \partial x^3 \equiv \partial\theta / \partial z = \{-1/\surd(1 - z^2/r^2)\}\{1/r - (z/r^2)\cos\theta\}$$

$$= (-1 + \cos^2\theta)/\surd(r^2 - z^2) = -(\sin\theta)/r,$$

$$\partial \bar{x}^3/ \partial x^1 \equiv \partial\varphi / \partial x = -y/x^2(1 + y^2/x^2) = -y/(x^2 + y^2)$$

$$= -r\sin\theta\sin\varphi/(r\sin\theta)^2 = -\sin\varphi / r\sin\theta,$$

$$\partial \bar{x}^3/ \partial x^2 \equiv \partial\varphi / \partial y = 1/x(1 + y^2/x^2) = x/(x^2 + y^2)$$

$$= r\sin\theta\cos\varphi/(r\sin\theta)^2 = \cos\varphi / r\sin\theta,$$

and

$$\partial \bar{x}^3/ \partial x^3 \equiv \partial\varphi / \partial z = 0.$$

A contravariant vector with components

$$A^1 = a, \qquad A^2 = b, \qquad A^3 = c \tag{5.21}$$

transforms under the coordinate transformation in Eqs. (5.19) according to the transformation law in Eq. (4.4.1b):

$$\bar{A}^{\alpha} = A^i (\partial \bar{x}^{\alpha} / \partial x^i) = A^1 (\partial \bar{x}^{\alpha} / \partial x^1) + A^2 (\partial \bar{x}^{\alpha} / \partial x^2) + A^3 (\partial \bar{x}^{\alpha} / \partial x^3).$$

Putting for the projection factors derived above and the components A^i in above equation, we obtain

$$\bar{A}^1 = (a \cos \varphi + b \sin \varphi) \sin \theta + c \cos \theta,$$

$$\bar{A}^2 = (1/r)\{(a \cos \varphi + b \sin \varphi) \cos \theta - c \sin \theta\},$$

and

$$\bar{A}^3 = (-a \sin \varphi + b \cos \varphi) / r \sin \theta. \ //$$

Example 5.3. A covariant vector has components

$$A_1 = x\,y, \qquad A_2 = 2y - z^2, \qquad A_3 = z\,x \tag{5.22}$$

in rectangular Cartesian coordinate system. Find its components in spherical polar coordinates.

Solution. Differentiating Eq. (5.1) we obtain the projection factors:

$$\left.\begin{array}{c} \partial x^1 / \partial \bar{x}^1 \equiv \partial x / \partial r = \sin \theta \cos \varphi, \ \partial x^1 / \partial \bar{x}^2 \equiv \partial x / \partial \theta = r \cos \theta \cos \varphi, \\ \partial x^1 / \partial \bar{x}^3 \equiv \partial x / \partial \varphi = -r \sin \theta \sin \varphi, \ \partial x^2 / \partial \bar{x}^1 \equiv \partial y / \partial r = \sin \theta \sin \varphi, \\ \partial x^2 / \partial \bar{x}^2 \equiv \partial y / \partial \theta = r \cos \theta \sin \varphi, \ \partial x^2 / \partial \bar{x}^3 \equiv \partial y / \partial \varphi = r \sin \theta \cos \varphi \\ \partial x^3 / \partial \bar{x}^1 \equiv \partial z / \partial r = \cos \theta, \ \partial x^3 / \partial \bar{x}^2 \equiv \partial z / \partial \theta = -r \sin \theta, \\ \partial x^3 / \partial \bar{x}^3 \equiv \partial z / \partial \varphi = 0. \end{array}\right\} \tag{5.23}$$

Applying the coordinate transformation law vide Eq. (4.5.1b) of a covariant vector given by Eq. (3.19) and putting from Eqs. (5.22) and (5.23) the components of the vector in the spherical polar coordinates are obtained:

$$\overline{A}_1 = xy \sin\theta \cos\varphi + (2y - z^2) \sin\theta \sin\varphi + zx \cos\theta$$
$$= r \sin\theta \{\sin\theta \sin\varphi (2 \sin\varphi + r \sin\theta \cos^2\varphi)$$
$$+ r \cos^2\theta (\cos\varphi - \sin\varphi)\},$$

$$\overline{A}_2 = r\{xy \cos\theta \cos\varphi + (2y - z^2) \cos\theta \sin\varphi - zx \sin\theta\}$$
$$= r^2 \cos\theta \{r \sin^2\theta \cos\varphi (\cos\varphi \sin\varphi - 1)$$
$$+ \sin\varphi (2 \sin\theta \sin\varphi - r \cos^2\theta)\},$$

$$\overline{A}_3 = r\{- xy \sin\theta \sin\varphi + (2y - z^2) \sin\theta \cos\varphi\}$$
$$= - r^2 \sin\theta \cos\varphi \{\sin\theta \sin\varphi (r \sin\theta \sin\varphi - 2) + r \cos^2\theta\},$$

where Eqs. (5.1) are also used. //

§ 6. $\boldsymbol{\nabla}$ operator in (r, θ, φ)

6.1. $\boldsymbol{\nabla} f$. There holds Eq. (5.9) for spherical polar coordinates. As a result, the expression for ∇f, given by Eq. (2.3b), reduces to

$$\nabla f = (\partial f / \partial r)\hat{\mathbf{t}}_r + (\partial f / \partial \theta)\hat{\mathbf{t}}_\theta / r + (\partial f / \partial \varphi)\hat{\mathbf{t}}_\varphi / r \sin\theta, \qquad (6.1a)$$

or, for Eq. (5.8)

$$\nabla f = (1/r)\Big\{\frac{\partial f}{\partial r}\cdot r \sin\theta \cos\varphi + \frac{\partial f}{\partial \theta}\cdot \cos\theta \cos\varphi - \frac{\sin\varphi}{\sin\theta}\cdot\frac{\partial f}{\partial \varphi},$$

$$(\partial f / \partial r) r \sin\theta \sin\varphi + (\partial f / \partial \theta) \cos\theta \sin\varphi + (\partial f / \partial \varphi) \cos\varphi / \sin\theta,$$

$$(\partial f / \partial r)\, r \cos\theta - (\partial f / \partial \theta) \sin\theta\Big\}. \qquad (6.1b)$$

6.2. $\boldsymbol{\nabla}.\mathbf{F}$. The divergence of a vector point function

$$\mathbf{F}(r, \theta, \varphi) \equiv f_1\, \hat{\mathbf{t}}_r + f_2\, \hat{\mathbf{t}}_\theta + f_3\, \hat{\mathbf{t}}_\varphi \qquad (6.2)$$

of spherical polar coordinates is evaluated from Eqs. (2.5) and (5.9):

$$\nabla.\mathbf{F} = \{\partial(r^2 f_1 \sin\theta)/\partial r + \partial(r f_2 \sin\theta)/\partial\theta + \partial(r f_3)/\partial\varphi\}/r^2 \sin\theta$$

$$=(1/r)\{2f_1 + r\,\partial f_1/\partial r + \partial f_2/\partial\theta + f_2 \cot\theta + (\partial f_3/\partial\varphi)/\sin\theta\}. \quad (6.3)$$

6.3. $\nabla^2 f$. Putting from Eq. (5.9) in Eq. (2.10), we derive the Laplacian of a scalar point function $f(r, \theta, \varphi)$:

$$\nabla^2 f = \frac{1}{r^2 \sin\theta}\left\{\frac{\partial}{\partial r}\left(\frac{\partial f}{\partial r} r^2 \sin\theta\right) + \frac{\partial}{\partial\theta}\left(\frac{\partial f}{\partial\theta}\sin\theta\right) + \frac{\partial}{\partial\varphi}\left(\frac{1}{\sin\theta}\frac{\partial f}{\partial\varphi}\right)\right\}$$

$$= \frac{\partial^2 f}{\partial r^2} + \frac{2}{r}\cdot\frac{\partial f}{\partial r} + \frac{1}{r^2}\left\{\frac{\partial^2 f}{\partial\theta^2} + \frac{\partial f}{\partial\theta}\cdot\cot\theta + \frac{\partial^2 f}{\partial\varphi^2}\cdot\operatorname{cosec}^2\theta\right\}. \quad (6.4)$$

6.4. $\nabla \times \mathbf{F}$. Putting from Eq. (5.9) in Eq. (2.12), we get the curl of the function defined by Eq. (6.2):

$$\nabla\times\mathbf{F} = \{\partial(r f_3 \sin\theta)/\partial\theta - \partial(r f_2)/\partial\varphi\}\,\hat{\mathbf{t}}_r/r^2\sin\theta$$

$$+\{\partial f_1/\partial\varphi - \partial(r f_3 \sin\theta)/\partial r\}\,\hat{\mathbf{t}}_\theta/r\sin\theta + \{\partial(r f_2)/\partial r - \partial f_1/\partial\theta\}\,\hat{\mathbf{t}}_\varphi/r \quad (6.5)$$

$$= (1/r)[\{f_3 \cot\theta + \partial f_3/\partial\theta - \operatorname{cosec}\theta(\partial f_2/\partial\varphi)\}\hat{\mathbf{t}}_r$$

$$+\{(\partial f_1/\partial\varphi)/\sin\theta - f_3 - r\partial f_3/\partial r\}\hat{\mathbf{t}}_\theta + \{f_2 + r\partial f_2/\partial r - \partial f_1/\partial\theta\}\hat{\mathbf{t}}_\varphi].$$

Example 6.1. For spherical polar coordinates

$$(\nabla\varphi)\times(\nabla\cos\theta) = (\nabla r)/r^2. \quad (6.6)$$

Solution. Applying Eq. (6.1a), we find

$$\nabla\varphi = \hat{\mathbf{t}}_\varphi/r\sin\theta, \quad \nabla\cos\theta = -(\hat{\mathbf{t}}_\theta\sin\theta)/r, \quad \nabla r = \hat{\mathbf{t}}_r. \quad (6.7)$$

Forming the cross product of above first two vectors, and putting from Eq. (5.11b), we obtain

$$(\nabla\varphi)\times(\nabla\cos\theta) = \hat{\mathbf{t}}_r/r^2,$$

which is the same as Eq. (6.6). //

Example 6.2. Evaluate $\mathbf{F} \times \text{curl}\,\mathbf{F}$, where

$$\mathbf{F} \equiv (r^2 \cos\theta)\,\hat{\mathbf{t}}_r - \hat{\mathbf{t}}_\theta / r + \hat{\mathbf{t}}_\varphi / r\sin\theta. \tag{6.8}$$

Solution. Comparing Eqs. (6.2) and (6.8), we note that

$$f_1 = r^2\cos\theta, \qquad f_2 = -1/r, \qquad f_3 = 1/r\sin\theta;$$

so that

$$\partial f_1/\partial\theta = -r^2\sin\theta,\ \partial f_1/\partial\varphi = 0,\ \partial f_2/\partial r = 1/r^2,\ \partial f_2/\partial\varphi = 0,$$

$$\partial f_3/\partial r = -1/r^2\sin\theta, \qquad \partial f_3/\partial\theta = -\cot\theta/r\sin\theta.$$

Therefore, Eq. (6.5) yields

$$\nabla\times\mathbf{F} = (r\sin\theta)\,\hat{\mathbf{t}}_\varphi. \tag{6.9}$$

Forming cross product of Eqs. (6.8) and (6.9), and using Eq. (5.11) we, therefore, get

$$\mathbf{F}\times\text{curl}\,\mathbf{F} = -r^3\,\hat{\mathbf{t}}_\theta\cos\theta\ \sin\theta - \hat{\mathbf{t}}_r\sin\theta$$

$$= -\sin\theta\,\{\hat{\mathbf{t}}_r + r^3\,\hat{\mathbf{t}}_\theta\cos\theta\}.\ //$$

CHAPTER 6

THEORY OF RELATIVITY

§ 1. Theory of relativity

The theory of relativity propounded by Albert Einstein includes two interrelated theories:

(i) special relativity, and (ii) general relativity.

Special relativity applies to all physical phenomena in the absence of gravity while the general theory of relativity explains the law of gravitation and its relation to other forces of nature. He found that space and time were interwoven into a single continuum known as space-time. Events occurring at the same time for one observer could occur at different times for another.

In the year 1905 A.D., Albert Einstein discovered that the laws of physics are the same for all non-accelerating observers, and the speed of light in a vacuum was independent of the motion of all observers. This was his theory of special relativity. It introduced a new framework for physics and proposed new concepts of space and time.

Einstein spent 10 years thereafter trying to include acceleration in the theory and published the theory of general relativity in 1915 A.D. He discovered that massive objects cause a distortion in space-time, which is felt as gravity. While working out the equations for his general theory of relativity, Einstein realized that massive objects caused a distortion in space-time.

Two objects exert a force of attraction on each other known as "gravity". While formulating his three laws of motion, Isaac Newton had quantified the gravity between two objects. The force exerted between two bodies depends on their masses and the distance between them. For example, the centre of the Earth pulls the body towards it keeping the body firmly on the ground. Simultaneously, the centre of mass of the body pulls back at the Earth. But the more massive body, i.e. the Earth hardly experiences the tug from the lighter body. Yet Newton's laws assume that gravity is an inherent force of an object acting over a distance.

1.1. Lorentz transformation

Considering

L = the length of a moving object, L_0 = its length in its rest position,

v = velocity of the moving object, c = velocity of the light,

M = the (relativistic) mass of the moving object,

and

M_0 = its invariant mass in its rest position,

the Lorentz transformations describe the following results:

$$L = L_0 \sqrt{(1 - v^2/c^2)}\,, \tag{1.1}$$

and

$$M = \frac{M_0}{\sqrt{(1 - v^2/c^2)}}\,. \tag{1.2}$$

Thus, when the moving object acquires the velocity of light, i.e. when v tends to c, L contracts to zero but the mass M becomes infinite.

BIBLIOGRAPHY

1. Eisenhart, L.P.: Riemannian Geometry. *Princeton Univ. Press, Princeton, N.J.* (*U.S.A.*), 6th Printing, 1966.

2. Kreyszig, E.: Advanced Engineering Mathematics. *John Wiley and Sons*, New York (*U.S.A.*), 7th ed., 1993.

3. Lipschutz, M.M.: Differential Geometry. *Schaum's Outline Series, McGraw-Hill Book Co., New York*, 1969.

4. Mishra, R.S.: A Course in Vectors and their Applications. *Prakashan Kendra, Lucknow* (*India*), 3rd ed., 1969.

5. Misra, R.B.: Tensors. *Hardwari Publications*, *Prayagraj* (*Allahabad*), *India*, 2002; MR 2003 d # 53022.

6. Misra, R.B.: Sadishon evam unke anuprayogon ka adhyayan (Hindi version of Prof. R.S. Mishra's above book). *Ibid*, 2002.

7. Misra, R.B.: A Text-book of Classical Mechanics. *Lambert Academic Publishers*, *Saarbrücken* (*Germany*), 2010, ISBN 978-3-8433-8306-6.

8. Misra, R.B.: Basic Mathematics at a Glance. *Ibid*, 2010, ISBN 978-3-8433-8696-8.

9. Misra, R.B.: Engineering Mathematics. *Ibid*, 2011, ISBN 978-3-8433-8931-0.

10. Misra, R.B.: Advanced Applied Mathematics. *Central West Publishing*, *Orange*, *NSW* (*Australia*), 2018, ISBN 978-1-925823-11-0.

11. Misra, R.B.: Mathematics for Engineers & Physicists - Pt. 1. *Ibid*, 2019, pp. xiv + 306, ISBN (print): 978-1-925823-51-6, ISBN (e-book): 978-1-925823-50-9.

12. Misra, R.B.: Glossary of Mathematical Terms and Concepts, Pt. 3. *Ibid*, 2019, ISBN (print): 978-1-925823-73-8.

13. Misra, R.B.: Glossary of Mathematical Terms and Concepts, Part 4, *ibid*, pp. xvi + 282, ISBN (print): 978-1-925823-74-5.

14. Weatherburn, C.E.: An Introduction to Riemannian geometry and the Tensor Calculus. *Cambridge University Press, Cambridge*, 1938, Paperback ed., December 2008; (Hindi translation by Prof. R. S. Mishra: *Uttar Pradesh Hindi Granth Academy, Lucknow, India*, 1971).

15. Willmore, T.J.: An introduction to differential geometry. *Oxford Univ. Press*, *London*, 1959; *Clarendon Press*, *Oxford*, 1972 (reprinted).

INDEX

CPSIA information can be obtained
at www.ICGtesting.com
Printed in the USA
LVHW081734170620
658145LV00014BB/1479

9 781925 823820